TECHNIQUES FOR IDENTIFYING TRANSURANIC SPECIATION IN AQUATIC ENVIRONMENTS

PANEL PROCEEDINGS SERIES

TECHNIQUES FOR IDENTIFYING TRANSURANIC SPECIATION IN AQUATIC ENVIRONMENTS

PROCEEDINGS OF A TECHNICAL COMMITTEE MEETING
ON THE BEHAVIOUR OF TRANSURANICS
IN THE AQUATIC ENVIRONMENT
AND SEDIMENT-WATER EXCHANGES:
TECHNIQUES FOR IDENTIFYING THE SPECIATION,
JOINTLY ORGANIZED BY THE
INTERNATIONAL ATOMIC ENERGY AGENCY
AND THE
COMMISSION OF THE EUROPEAN COMMUNITIES
AND HELD IN ISPRA, ITALY, 24–28 MARCH 1980

INTERNATIONAL ATOMIC ENERGY AGENCY
VIENNA, 1981

TECHNIQUES FOR IDENTIFYING TRANSURANIC SPECIATION
IN AQUATIC ENVIRONMENTS, IAEA, VIENNA, 1981
STI/PUB/613
ISBN 92−0−021081−3

Printed by the IAEA in Austria
December 1981

FOREWORD

To establish reasonable discharge limits for disposal of radioactive wastes to the aquatic environment, and to make meaningful assessments of environmental impacts, it is important to improve techniques for measuring radioactivity in various environmental compartments. A considerable range of methods has been used for these analyses by investigators both in the field and in the laboratory. Because of the variety of approaches that can be used to gather the basic data needed for release rate determination the results of individual investigators are often difficult to compare. Recently considerable progress has been made in improving techniques for measuring radioactivity in marine organisms and their environment. These techniques, for application to either in-situ or laboratory studies related to the assessment of the behaviour of transuranics and other stable and radioactive elements, require careful consideration in order that their importance, applicability and limitations are well understood.

For this reason the International Atomic Energy Agency and the Commission of the European Communities sponsored a meeting of experts in radioecology to consider the problems involved when an attempt is made to detect the total concentrations of long-lived radioisotopes, and to determine their physical and chemical forms and their subsequent availability to marine organisms.

Thus, a joint CEC/IAEA technical meeting was held in Ispra on 24–28 March 1980 to consider the following specific subject areas: methods of determining transuranic chemical species at environmental levels of $\leqslant 10^{-10}M$; methods of studying the bioavailability of transuranics in aquatic organisms both in laboratory and in-situ investigations; and chemical methods of determining metal fractions applicable to transuranic analysis, on fresh, estuarine and coastal sediments.

The present report gives recommendations on methodology useful for carrying out reliable quantitative measurements, and provides for reference analytical methods for the determination of selected radionuclides including the collection, storage and preparation of samples for chemical and radiochemical analyses. It is hoped that the information will be helpful to radioecologists undertaking similar studies and will contribute to the production of not only reliable data but useful data in radionuclide measurements on environmental samples, thus in turn contributing to the safeguarding of the quality of human life.

CONTENTS

**METHODS OF DETERMINING TRANSURANIC CHEMICAL
SPECIATION AT ENVIRONMENTAL LEVELS $\leqslant 10^{-10}$ M (Session I)**

METHODS OF STUDYING THE BIOAVAILABILITY OF TRANSURANICS IN AQUATIC ORGANISMS (Session II)

CHEMICAL METHODS OF DETERMINING METAL FRACTIONS AND APPLICABILITY TO TRANSURANICS ON FRESH, ESTUARINE AND COASTAL SEDIMENTS (Session III)

Session I

METHODS OF DETERMINING
TRANSURANIC CHEMICAL SPECIATION
AT ENVIRONMENTAL LEVELS $\leqslant 10^{-10}$M

Scientific Secretary

W. FORSTER
IAEA, Vienna

Chairmen

D.N. EDGINGTON
United States of America

E. HOLM
Sweden

CHARACTERIZATION OF TRANSURANIC ELEMENTS AT ENVIRONMENTAL LEVELS

D.N. EDGINGTON
Centre for Great Lake Studies,
University of Wisconsin-Milwaukee,
Milwaukee, Wisconsin,
United States of America

Abstract

CHARACTERIZATION OF TRANSURANIC ELEMENTS AT ENVIRONMENTAL LEVELS.
This paper summarizes current understanding of transuranic speciation in natural waters.
Several predictive models describing the behaviour of these elements in both marine and fresh-
water systems illustrate the importance of this information.

1.   INTRODUCTION

The purpose of this paper is to summarize what we know concerning the
measurement of transuranic elements in natural waters in terms of chemical
speciation and the importance of this information in constructing predictive
models for the behaviour of these elements in marine and freshwater environments
leading to their transfer to man. With recent developments in instrumentation for
low (essentially zero) background alpha spectroscopy, scientists investigating the
behaviour of elements in the environment have a technique which is unique in
that they can study the behaviour of the transuranic elements not only at extremely
low concentrations ($\sim 10^{-18}$M or 10^6 atoms/litre) but also in the absence of any
contamination or natural background. The latter is very important in that the
transuranic elements will not be in equilibrium, as are many natural elements,
in the environment and therefore may be used as tracers for rate studies.

The results of these experiments will also provide a model for attempts to
describe the behaviour and speciation of other elements in the aquatic environment, particularly in situations where it is believed that the system is approaching
steady-state or chemical equilibrium.

Since almost all heterogeneous processes in the water column involve transfer
between a species in solution and its association with another phase, be it a sediment particle, organic detritus, or a living organism, it is clear that a knowledge
of the speciation of the ions in aqueous solution is critical to our understanding
of the adsorption or absorption process at the interface. The range of interactive
reactions, physical, chemical, and biological, involved in the aquatic system is
summarized in Fig.1. While the change in concentration of radioactivity in the

3

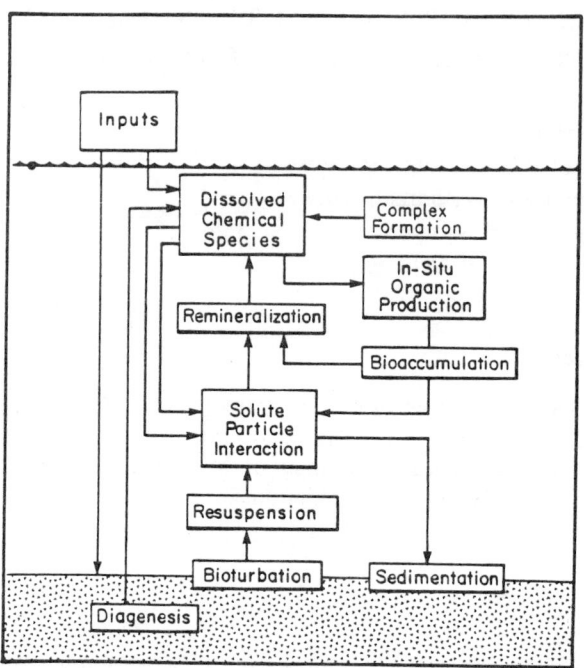

FIG.1.   *Schematic representation of the possible interactions of transuranic elements with particles, biota, and sedimentation in the oceans.*

water column with time can be described, for certain large bodies of water, in terms of a fairly simple time-concentration model involving an assessment of inputs, outputs, residence times of particles and water in the prescribed basin [1], such simple models will break down when the rate of addition of new inputs approaches zero and the return of material to the water column by desorption reactions, such as the decomposition of organic detritus or mass action reactions, becomes dominant. Thus it is essential, in order to fully understand the long-term behaviour of pollutants in the marine environment, that experiments are conducted to assess these important steady-state or equilibrium conditions. In fact, for very long-lived radionuclides, such as plutonium, a complete assessment of simple long-term equilibrium models will be far more important than kinetic models with complicated differential equations. However, in order to interpret measurements under steady-state or equilibrium conditions and to have the ability to predict the effects of changing conditions of water quality or composition, it is imperative that we have an understanding of the chemical properties of these elements under widely differing environmental conditions.

The study of the chemical speciation of the transuranic elements at their present concentrations in natural waters is not a simple problem. Speciation will be affected by hydrolysis, complexation, redox reactions, and at higher concentrations by polymer formation and disproportionation reactions [2]. Therefore many of the techniques which are available for the study of speciation of other metals, e.g. spectrophotometry, potentiometric titration, polarography, etc., cannot be used to characterize the transuranic elements under environmental conditions at concentrations where these techniques are effective. This is because of the strong tendency of tetravalent ions to undergo hydrolysis to polymeric hydroxo-complexes.

2.   SOURCES

It has been suggested that the behaviour of transuranic elements in the environment is a function of source term and, for plutonium, isotopic composition. The major source of most of the transuranic elements in the environment is from the testing of nuclear weapons. As a result of these tests radioactive debris was introduced into the atmosphere which then returned slowly to the earth's surface. Until recently it has been assumed that any transuranic element returning to earth would be, as a result of exposure to extremely high temperature in the explosion, in the form of high-fired oxide. These oxides are known to be very refractory and would therefore be insoluble in natural waters. Even if these particles did dissolve, insoluble polymeric hydroxides would form. In either case the material would be unavailable for biological uptake, other than by filter feeders. While it may be argued that large fallout particles have been collected, there is no evidence to suggest that these could be pure high-fired plutonium oxide particles. In fact, most of the plutonium returning to earth was formed by radioactive decay of ^{239}Np a few days after the explosion of the weapon [3]. Extensive measurements of plutonium in large numbers of water samples have shown no evidence of extreme variability which would arise from the random collection of hot particles. However, there is evidence of hot particles from the analysis of samples from the area around Greenland which was contaminated by the accidental dispersal of (nuclearly) unexploded plutonium [4].

Transuranic elements have also been released to the environment during fuel reprocessing and fabrication activities. The chemical form of these releases may range from relatively insoluble oxide particles to relatively soluble inorganic salts and organic complexes which may be present in solid and liquid wastes.

A summary of the major sources of transuranic elements in the environment with an estimate of the magnitude of the total released or the total released per year is given in Table I. It appears that, with the possible exception of that material discharged into Bombay Harbour, plutonium appears to behave in a very consistent manner in all marine environments as well as in large freshwater lakes such as the Great Lakes [5]. However, as will be illustrated later, in small

TABLE I. STUDIES OF TRANSURANIC ELEMENTS IN MARINE AND COASTAL ENVIRONMENTS

Location	Aquatic system	Source and nuclides studied	Inventory
Oceans	Atlantic Pacific	Atmospheric fallout: ^{239,240}Pu, ^{238}Pu, ^{241}Am, and ^{241}Pu	~150 kCi Pu
Greenland (Denmark)		Atmospheric fallout Accidental contamination, Pu	$\geq$20 Ci Pu
Coastal zones of the USA	Hudson River	Atmospheric fallout: ^{239,240}Pu, ^{238}Pu, ^{241}Am, and ^{241}Pu	Annual transport, about 6 mCi
	Savannah River	Effluents from processing plants: ^{238,239,240,241}Pu and ^{241}Am	Fallout +0.3 Ci from SRP
India	Bombay	Effluents from processing plants: ^{238,239,240,241}Pu and ^{241}Am	?
France	English Channel Cap de la Hague	Effluents from processing plants: ^{238,239,240,241}Pu and ^{241}Am	25 Ci total Pu
United Kingdom	Irish Sea (Windscale)	Effluents from processing plants: ^{238,239,240,241}Pu and ^{241}Am	Has been up to 100 Ci/month for Pu and >200 Ci/month for Am
UK and FRG	North Sea	Atmospheric fallout and reprocessing wastes from the Irish Sea and Cap de la Hague	
Monaco	Mediterranean	Atmospheric fallout	

Location	Aquatic system	Source and nuclides studied	Inventory
Test sites at Enewetak	Tropical Lagoon	Debris from weapons testing and immediate fallout: ^{238,239,240}Pu, ^{241}Pu, ^{241}Am, and possibly ^{237}Np and the higher actinides	1200 Ci ^{230,240}Pu 475 Ci ^{241}Am
Bikini	Tropical Lagoon	Ibid.	1470 Ci ^{239,240}Pu 1140 Ci ^{241}Am
Great Lakes in the USA	L. Michigan L. Huron L. Erie L. Ontario	Atmospheric fallout: ^{239,240}Pu, ^{238}Pu, ^{241}Am, ^{241}Pu	~120 Ci Pu ~60 Ci Pu

eutrophic lakes, lakes with high contents of carbonate or sulphate ions, or Bombay Harbour (which may be a good example of a eutrophic marine environment), the plutonium appears to behave very differently [6]. Studies of the behaviour of transuranic elements in areas such as highly saline lakes, bogs, or even the African rift lakes, with high concentrations of organic carbon, carbonate, sulphate and other ions will be very important in assessing the possible behaviour of these elements in ground water.

3.   ANALOGUES, ISOTOPES, AND OXIDATION STATES

The transuranic elements are a subset of the actinide series which is similar to the lanthanide series in that electrons are added successively to the 5f orbitals rather than the 4f orbitals. Since the 5f electrons are less effectively shielded than the 4f electrons, the actinide elements have more complex chemical properties than the lanthanides. In particular, the oxidation reduction behaviour of uranium, neptunium, and plutonium is complicated by the presence of multiple oxidation states in solution. However, americium and curium appear to behave similarly to the rare earths and probably can exist only in the trivalent state in the environment.

A summary, for each element, of the oxidation states that can exist in the environment, together with possible analogue elements (stable or radioactive) are given in Table II. Because these analogue elements are likely to carry through in group separation schemes, values for the half-life of each radioactive isotope and the principal alpha energies are also given.

Because of its similar chemical properties, ^{227}Ac is likely to be carried with americium and curium. The short-lived daughter products of ^{227}Ac which grow in rapidly after separation have alpha energies which can seriously interfere in the detection of both ^{242}Cu and ^{244}Cu. They are ^{227}Th and ^{223}Ra with principal alpha energies of 6.04 and 5.72 MeV, respectively. Fortunately, the chemical properties of actinium and curium are sufficiently different so that effective separations can be made [7]. However, care must be taken to make sure that an efficient separation is made.

The obvious analogues for the tetravalent transuranic elements are the naturally occurring isotopes of thorium, ^{228}Th, ^{230}Th, and ^{232}Th. There is little evidence for uranium existing in the present aquatic environment as U(IV). The tetravalent transuranic elements include plutonium and possibly neptunium and americium. In measuring ^{238}Pu, particularly in sediments, contamination of the final counting disc with ^{228}Th is a major problem. ^{238}Pu and ^{228}Th have very similar alpha energies (5.50 and 5.42 MeV, respectively). Thorium has a tendency to tail in the ion-exchange separation procedures used for the separation of plutonium and, if the separation is not carefully carried out, the ^{238}Pu will be

TABLE II. ISOTOPES, OXIDATION STATES, ANALOGUES, AND ALPHA ENERGIES

	III		IV	V		VI
	Rare earths		^{228}Th	^{231}Pa		^{234}U
			1.9 a	3.4×10^4 a		2.4×10^5 a
		(B)	5.42 MeV	5.01 MeV	(C)	4.78 MeV
	^{227}Ac		^{230}Th			^{235}U
	21.8 a		7.7×10^4 a			$7. \times 10^8$ a
	4.94 MeV		4.69 MeV			4.37 MeV
	(^{227}Th)					
(A)	5.75 MeV		^{232}Th			^{238}U
	4.94 MeV		1.4×10^{10} a			4.5×10^9 a
			4.01 MeV			4.20 MeV
	(^{223}Ra)					
(A)	5.72 MeV		Np	^{237}Np		
	5.61 MeV			2×10^6 a		
				(C) 4.79 MeV		
	Pu		^{238}Pu	Pu		Pu
			88 a			
		(B)	5.50 MeV			
			^{239,240}Pu			
			2.4×10^4 a			
			6500 a			
			5.16 MeV			
			5.17 MeV			
	^{241}Am		Am?			
	433 a					
	5.49 MeV					
	^{242}Cm					
	163 d					
	6.11 MeV					
	6.07 MeV					
	^{244}Cm					
	18 a					
(A)	5.80 MeV					
	5.76 MeV					

Note: Daughter products are shown in parentheses. Interfering pairs of alpha energies are labelled (A), (B) and (C).

found as a shoulder on the ^{228}Th peak as a result of the concentrations of ^{228}Th in sediment samples being two or three orders higher than that of ^{238}Pu.

For the pentavalent transuranic elements, the natural analogue should be protactinium. However, the chemical properties of neptunium, whose pre-dominant oxidation state is (V) are very different from those of protactinium [7]. While neptunium (V) appears to behave like a simple monovalent ion in solution, protactinium has hydrolysis properties more reminiscent of thorium or plutonium (IV). The most recent experimental evidence suggests that a major fraction of plutonium is present in ocean water as Pu(V) [8].

The obvious analogue for the hexavalent transuranic elements is uranium(VI). Many studies have been made of the speciation of uranium in ocean water. The natural isotopes are ^{234}U, ^{235}U, and ^{238}U. Of the transuranic elements only Pu(VI) might be expected to be found in the environment. However, it should be pointed out that the alpha energies of ^{234}U and ^{237}Np are almost identical and, therefore, considerable care must be taken when attempting to measure ^{237}Np in environmental samples to ensure complete separation of neptunium from uranium [9].

4. MEASUREMENT OF CONCENTRATIONS

The measurement of the concentrations of transuranic elements in environ-mental samples, even though these concentrations are extremely low (10^{-18} to 10^{-10}M), is made easier in that the possibility of contaminating the sample during collection and analysis is small because the natural background is zero. However, there is some evidence that some reagents are sufficiently contaminated with plutonium to warrant a careful evaluation of blanks if reproducible values are to be obtained. It has been shown that these problems can be minimized if con-centrated stock solutions of all reagents are carefully filtered before dilution and use [8].

A generalized flow diagram for the measurement of transuranic elements in environmental samples is given in Fig.2. Since the concentration of these elements in most water samples is extremely low, an efficient preconcentration step is essential. In general this is accomplished by coprecipitating these elements onto iron hydroxide [11], calcium fluoride [12], calcium oxalate [13], and manganese oxide [14]. These methods have been reviewed previously [15]. Before pre-cipitation, a known concentration of a different isotope of each element to be determined is added to act as a yield monitor. This is, in effect, an isotope diluent. Therefore, it is extremely important that steps are taken to ensure that complete isotopic exchange occurs before forming the precipitate. If plutonium is being measured steps must be taken to ensure the oxidation state is adjusted. This is normally accomplished using a bisulphite reduction step. For example, it is now known that plutonium in the Irish Sea water is predominantly present as

MEASUREMENT OF CONCENTRATIONS

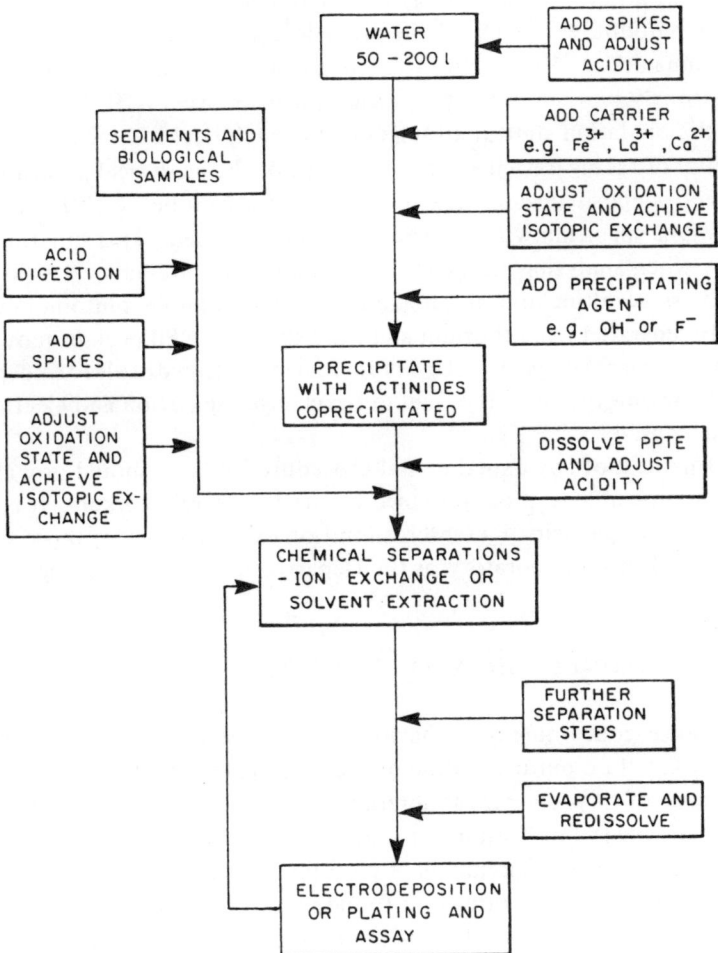

FIG.2.   Steps in the measurement of transuranic elements in natural samples.

Pu(V), while the ^{236}Pu or ^{242}Pu added will be present as Pu(IV). For reliable results the correct adjustment of the oxidation state is most critical.

In carrying out the preconcentration step, as well as in later steps in the chemical separation and purification [15], it is important that the greatest possible recovery of the activity is achieved. If low recoveries are encountered, the validity of the results could be questionable and, therefore, a careful evaluation should be made of all steps in the analytical procedure. It should be a standard procedure to keep all filtrates, ion exchange material, and glassware until the final planchette has been counted and its radiochemical purity evaluated. In this manner an important sample is rarely completely lost.

When in doubt, every step of the procedure should be tested using a tracer such as gamma-active ^{237}Pu. Since the separation of the various actinide elements depends strongly on their redox properties, the effectiveness of the methods used to adjust the oxidation state at all stages of the analytical scheme is critical.

In spite of taking extreme care at all steps of the analytical procedure, the sample electrodeposited onto the counting disc may not be weightless and a degraded alpha spectrum with poor resolution is obtained. This situation occurs commonly when analysing soil or sediment samples. The final deposit is either not weightless, resulting in peak broadening or, for example, plutonium is not completely separated from thorium and the expected ^{238}Pu peak cannot be resolved from the ^{228}Th peak. The simple remedy is to redissolve the deposit from the plate and to repeat the final ion-exchange separation and electro-deposition steps.

The final test of any experimental procedure is that it should provide results which are comparable to those obtained by other research workers. It is therefore important to participate in intercalibration exercises such as those sponsored by the IAEA Monaco Laboratory or the Department of Energy in the USA.

5.   DEVELOPMENT OF IDEAS OF SPECIATION

Early ideas concerning the behaviour of plutonium in aquatic environments were largely based on our perception of the chemical form of Pu in fallout and our knowledge of its properties as determined in the laboratory such as the stability of the Pu(IV) oxidation state and the formation of intractible colloids at pH $\geqslant 2$. However, measurements of plutonium in large bodies of water have demonstrated that the concentrations found in multiple samples collected at the same time and place are very self consistent. This would not be the case if the Pu was present in discrete fallout particles or colloids. The concentrations are also far higher than would be expected from the solubility of the Pu(IV) hydroxide but lower than for the solubility of the Pu(VI) hydroxide.

5.1. Fractionation of water samples

It is possible to measure plutonium, americium, etc. in the environment at concentrations as low as 10^{-18}M in the Great Lakes [16], a little higher in the open oceans [17], and about three orders of magnitude higher in the Irish Sea at the point of discharge from Windscale [13]. What can be learned about speciation of this element in natural waters from these measurements? The scheme shown in Fig.3 was developed a few years ago to examine the chemical form of plutonium in the Great Lakes [18]. The water samples are divided into a series of fractions

FRESHWATER FRACTIONATION SCHEME

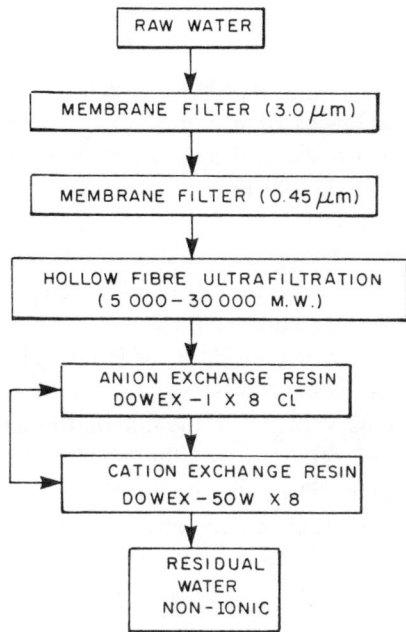

FIG.3.   Flow-chart for the determination of the predominant physical compartments containing transuranic elements in natural waters.

using physical and chemical techniques which could be adapted to handling the large volume of sample required. This scheme could be used for ocean waters; similar ion-exchange techniques have been used to study the speciation of plutonium in the waters of Bombay Harbour [6].

The first stage is to filter the raw water sample. Typically between 10 and 20% of the total Pu in the sample is collected on the filter [19]. In terms of an estimate of the total soluble plutonium in the ocean, the difference between measurement of filtered and unfiltered samples is unimportant. However, the measure of the actual concentration of plutonium in the particulate matter captured on the filter is very important (see below).

At this stage it is clear that the majority of the plutonium is in the solution phase. But, is it in true solution as simple ions or complexes, or is it, as has been suggested in many papers, in a colloidal form?   Such forms must result from complexing with naturally occurring organic or inorganic colloids, since at extremely low concentrations of Pu in natural waters (10^{-8} to 10^{-14}M) the probability of the formation of $-(Pu-O-Pu)_n-$ polymers is extremely unlikely.

At these concentrations, even the formation of dimers or other two-atom reactions such as disproportionation are unlikely. (A concentration of 1 fCi ^{239,240}Pu per litre is equivalent to $\sim 10^{-17}$M or ~ 6000 atoms·ml^{-1}, or 1 atom of Pu per 5×10^{18} molecules of water !)

Experiments carried out using large hollow fiber ultra-filtration devices with molecular weight cutoffs as low as 30 000 or 5000 molecular weight (MW) have shown that in both the Great Lakes and Irish Sea there is evidence of association of a small but real fraction of the plutonium with natural colloids as might be expected in these waters [10, 18]. Such waters have a relatively low content of dissolved organic matter and may be classed as oligotrophic. In other waters, where there are extremely high concentrations of organic matter, the formation of complexes with natural organic complexing agents is very important having a large effect on redox properties of plutonium (vide infra) [20]. Other experiments conducted at Bombay using low-level wastes added to sea water, containing relatively high concentrations of dissolved organic matter and resulting in dissolved plutonium concentrations of $\sim 10^{-11}$ to 10^{-10}M indicate the preferential formation of non-ionic species with time. The experiments also showed that the relative concentration of anionic to cationic species also increases with time. However, a large proportion of the added plutonium precipitated out [6].

5.2. Oxidation state

The behaviour of neptunium and, in particular, plutonium in the aquatic environment will be strongly dependent on the dominant oxidation state or ratio of oxidation states that can exist under the widely differing chemical properties of the receiving waters. For example, Pu(III) will behave like the rare earths, Pu(IV) or Np(IV) like Th(IV), and Pu(VI) like U(VI). As early as 1972 several authors had considered the complex interactions between the various oxidation states of Np and Pu in natural waters [21, 22]. The behaviour of Np and Pu will be governed in part by their total concentration in solution. At large enough concentrations disproportionation reactions between oxidation states are common [7]. However, under the conditions likely to apply to the release of these elements to the environment, the stable oxidation state(s) will be a function of pH, the presence of oxidizing or reducing agents, and complexing ligands.

Standard oxidation-reduction potentials may be used to predict the possible oxidation states of these elements in solution under environmental conditions. Such predictions are for the thermodynamic stability of each species and do not consider the effects of reaction rate as a function of concentration. Because measurements of the potentials for the four oxidation states of plutonium have not been made in near neutral solution, their magnitude must be estimated from measurements made in acid solution and values of the stability constants for the formation of hydroxyl and other complexes, which are not well known.

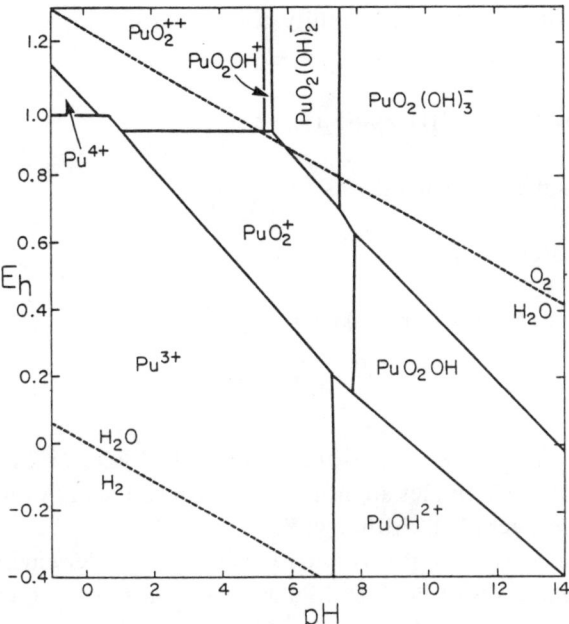

FIG.4. E_h-pH diagram for the speciation of plutonium in equilibrium with PuO_2 in water. (Data taken from Ref.[24].)

Several authors have used potentials calculated in this manner to construct E_h-pH diagrams similar to the one shown in Fig.4 [23]. This diagram delineates conditions where equal concentrations of pairs of species co-exist. The concentration of any species is assumed to be controlled by solid PuO_2 [21]. As early as 1972 it had been predicted that the predominant form of plutonium in natural waters would be Pu(V) [16]. Because of the tendency of highly charged cations to hydrolyse, free Pu^{2+} can exist only under strongly oxidizing and acid conditions. In the normal pH range of natural waters, 6–8, the results of these calculations suggest that in normal oxygen-rich waters, where the E_h value is approximately 0.5 V [24], plutonium will be present predominantly as Pu(V), PuO_2^+. Under more strongly alkaline conditions (pH > 8), the hydroxyl complex of Pu(VI), $PuO_2(OH)_3^-$ is stable enough to possibly become the predominant species rather than PuO_2^+. In lakes which are more acid with a pH < 6, or more reducing with an E_h < 0.5 V, Pu(III) as Pu^{3+} should become more dominant.

The effect of hydrolysis on the relative proportions of the various oxidation states is also illustrated in Fig.4. Furthermore the formation of complexes can also stabilize different oxidation states in solution. It is well known that, in general, the order of magnitude of stability constants for complexes with the actinide elements varies as

$$M(IV) > MO_2(II) > M(III) > MO_2(I),$$

for divalent anions it varies generally as

$$CO_3^{2-} > oxalate^{2-} > SO_4^{2-}$$

and for monovalent anions it varies as

$$F^- > NO_3^- > Cl^- > ClO_4^-$$

Unfortunately, the stability constants for many of the hydroxo, carbonato, and other complexes of the actinides are not well known and many values given in the literature are suspect [27]. Although Pu^{4+} tends to form the strongest complexes, the concentration of this free ion is so small under environmental conditions that complexes of Pu^{3+} and PuO_2^{2+} may contribute a greater proportion to the total concentration of plutonium in solution, assuming that complexes of PuO_2^+ are unlikely.

Similar E_h-pH diagrams can also be constructed for americium. While there is no argument that Am(IV) can exist in strongly alkaline solutions, there is some uncertainty in the literature regarding the exact value of the redox potential for the oxidation of Am(III) to Am(IV) under alkaline conditions [25, 26]. Therefore it is possible that Am(IV) could possibly exist in strongly-oxidizing natural waters.

Over the years many techniques have been developed to separate actinide elements in different oxidation states from one another [7]. Recently one of the original coprecipitation techniques developed by Seaborg and coworkers for the separation of plutonium [7] has been adapted for use with large volume water samples [28]. The technique depends on the simple observation that only actinides in the (III) and (IV) oxidation states are coprecipitated with rare earth fluorides. Several other compounds such as zirconium phosphate and bismuth phosphate have also been used to carry out this separation. Using these techniques it has been found that the plutonium in the waters of the Great Lakes, Irish Sea, and Enewetak Lagoon are predominantly in one of the two upper odixation states, the ratio of Pu(V and VI)/Pu(III and IV) varying between 4 and 10 [28]. However, in the interstitial waters of sediments from the Irish Sea [29] and in natural waters with high concentrations of organic carbon complexing agents such as SO_4^{-2}, Pu(III and IV) predominates [30].

More recently, a laboratory technique developed to separate Np(IV), (V), and (VI) using silica gel [31] has also been adapted to be used with large volume

environmental samples [8]. As a result it has been shown that, in the waters of
the Great Lakes and the Irish Sea, plutonium is present as Pu(V) and not as
Pu(VI). If Pu is added as Pu(VI) it is rapidly transformed into Pu(V). Pu(V) has
also been characterized in pond waters on the basis of adsorption behaviour of
plutonium in different oxidation states on TiO_2 [32].

5.3. Association with sediments

Even though it is relatively simple to measure the concentration of the
transuranic elements and other long-lived radionuclides in the water column,
a large proportion of the activity has become attached to sediment particles and,
in shallow basins such as the Great Lakes and the Irish Sea, was transferred
rapidly to the sediments [13, 33]. Measurements of the change in concentration
with depth in the sediments can provide information on the history of deposition,
sedimentation rates, mixing processes within the sediments, and particle transport
in the water column [33, 34]. Measurements of the distribution of the radioactivity associated with different geochemically defined phases such as specific
minerals or surfaces in the sediment can provide information on the possible
chemical reactions controlling uptake by the sediments and may provide insight
into the mechanisms controlling the availability of trace contaminants to
organisms [35, 36]. Changes in concentration or distribution between phases
of radionuclides can also provide insight into changes in behaviour due to
diagenetic reactions such as the formation of complexing agents or changing
the surface properties of the particles, and possible changes in the effectiveness
of sediments to act as sinks for these and similar contaminants.

Many simple extraction techniques have been devised to determine the
association of trace contaminants with operationally defined geochemical
fractions and these are discussed in detail in another paper in this volume [37].
However, a simple scheme which was used to investigate transport phases in
suspended river sediments has been used to investigate the geochemical
associations of Pu, ^{137}Cs, and ^{210}Pb in lacustrine and nearshore sediments [35].
The separation scheme used in these experiments, the geochemical phases
extracted, as well as the results for sediments from Lake Michigan are shown in
Table III. The experimental details may be found elsewhere [36]. The order of
extraction with different reagents could lead to significant differences in the
fractionation of different elements. However, for Pu and ^{137}Cs, changing the
order of extraction had little effect on the results shown in Table III. For
sediments from the Great Lakes, Irish Sea, or Buzzards Bay, plutonium and
americium are found only associated with the reducible oxides and this association
does not change with depth in the sediment column [35, 36]. The strong binding
of ^{137}Cs with lattice sites in clays like illite is responsible for the strong irreversible
association of this nuclide with sediments in the Great Lakes. Hence the occurrence

TABLE III. TREATMENT OF SEDIMENTS WITH SELECTIVE EXTRACTION
AGENTS

Phase	Reagent	Pu	^{137}Cs	^{210}Pb
Reducible hydroxides, carbonates etc.	Dilute acid	+	0	+
Ion exchange	0.1M $MgCl_2$	0	0	0
Reducible oxides and hydroxides	Na-citrate/- Na-dithionite	+	0	X
Organics	NaOH	0	0	X
Carbonates	Na-acetate	0	0	X
?	Dilute acid	0	0	0
Crystalline lattice	Fusion	0	+	−

(+)   Complete recovery of activity ($\geqslant$90%).   (X) Incomplete recovery of activity ($5 \leqslant x \leqslant 90\%$).
(0)   No recovery ($\leqslant$5%). (−) Not used.

of ^{137}Cs in the crystalline lattice fraction is not surprising. It might be expected
that the ^{137}Cs associated with marine sediments will be more mobile and appear
in the ion-exchangeable fraction.

In contrast, the behaviour of ^{210}Pb is more complicated: its distribution is
approximately equal between the three extractible phases − reducible hydrous
oxides, organic and carbonate fractions − but still half remains which is returned
to solution in dilute acid [38]. Furthermore, the distribution of ^{210}Pb appears
to change with depth in the sediment core. Thus the behaviour of two heavy
elements in the sediments is very different even though the major source of
each is the transfer of single atoms on aerosols from the air to the water surface,
association with particles in the water column, and finally transfer to the sediment
column.

Finally, measurement of the oxidation state of plutonium in solution after
extraction of sediment (from the Great Lakes or Irish Sea) with dilute acid in
the presence of a holding oxidant has indicated that all of the plutonium on
particles is present as Pu(III) or (IV). If higher oxidation states are present, their
concentration is small and not detectable within the accuracy of the experiment [28].

6.   REACTION MECHANISMS

The association of plutonium or ^{137}Cs with a single phase in the sediments
suggests that the underlying mechanisms for these reactions may be comparatively
simple. This leads to the hypothesis that the behaviour of plutonium and ^{137}Cs

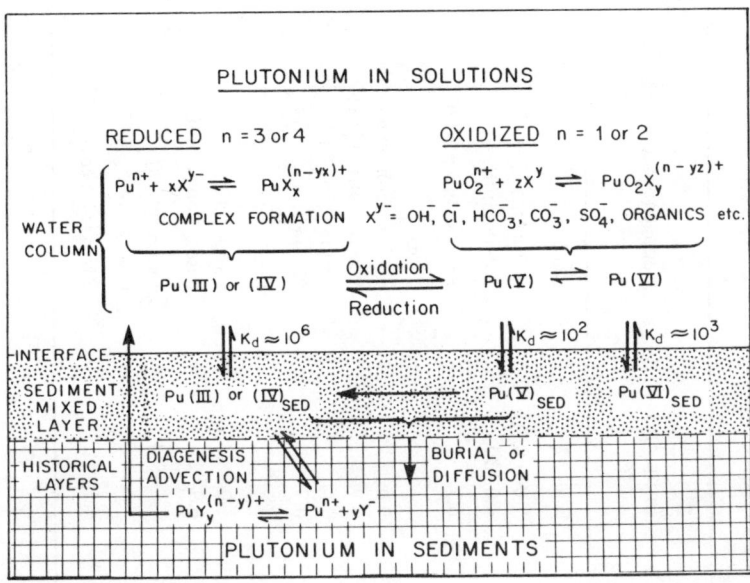

FIG.5. *A model of the chemical processes involving plutonium in natural waters.*

in aquatic environments could be interpreted in terms of simple adsorption or ion-exchange mechanisms. Such a mechanism is not unreasonable and it has been already suggested that the concentrations of trace metals in the oceans are controlled by an equilibrium distribution between the dissolved and solid phases. The reaction scheme for plutonium is complicated by the observation regarding changes in the oxidation state between the water and the sediments. These reactions can be summarized in terms of three equilibrium reactions which are illustrated in Fig.5. The equilibria are: (1) the redox equilibrium between oxidation states in the water column which can be strongly affected by (2) the formation of complexes with OH^-, HCO_3^-, CO_3^{2-}, SO_4^{2-}, Cl^- and other inorganic or organic ligands, and (3) the reaction of these complexes with the iron-manganese surface layers of the sediments.

There is a considerable amount of data available to suggest that these reactions are probably the same in the oceans and Great Lakes, but that changes in water chemistry can strongly affect the position of all three equilibria. As noted above, the ratio of Pu(V)/Pu(III or IV) is approximately the same and favouring Pu(V) in ocean or Great Lakes water (high pH and carbonate, low dissolved organic carbon), but strongly favours Pu(III or IV) in organic carbon rich waters [30]. However, the concentration of Pu(V) remains essentially constant in the latter water samples.

TABLE IV. DISTRIBUTION COEFFICIENTS, K_d, FOR PLUTONIUM IN DIFFERENT NATURAL WATERS

Sampling area	Concentration on particles (pCi·g^{-1})	Concentration in solution (pCi·ltr^{-1})	K_d (ml·g^{-1}) $\times 10^4$	Comments and References
Bombay Harbour	0.4–2.9	4.0–20	4.8–13	Silt [6]
Enewetak Lagoon	0.1–75	2.0–75	5.0–60	Sediments [39]
			4.0–30	Lab.desorption
Hudson River	0.02	0.3	6.7	River sediments [20]
Irish Sea	36	500	7.5	1 km ⎫
	23–80	370–650	4.6–21	4.5 km ⎬ from Windscale
	9.0–29	310–460	1.9–9.3	9 km ⎬ outfall [13]
	0.22	35	0.6	75 km ⎬
	0.37	50	0.7	90 km ⎭
Lake Michigan	0.2	0.6	33	[36]
Lake Washington	–	–	1.0–8.5	Lab.sorption/
	–	–	6.9–27	desorption　　[40]
Mediterranean Sea	–	–	1.6–9.4	Lab.sorption/
				desorption　　[41]
Savannah River	0.014–0.1	0.24–2.4	4.2–41	River sediments [20]

From measurements of plutonium in surface sediments and overlying water it is possible to calculate a constant, often called a distribution coefficient viz.

$$K_d = \frac{\text{concentration of radionuclide in dry sediment } (fCi \cdot g^{-1})}{\text{concentration of radionuclide in solution } (fCi \cdot ml^{-1})}$$

Values of this coefficient calculated for measurements made in many environments are shown in Table IV. The majority of values are in the range of $1 \times 10^4 - 5 \times 10^5$. Since variations of $5-10$ might be expected on the basis of particle size, then considering the wide variety of sediment types and possible differences in their particle size distribution, this relatively small range in values is striking and suggestive that a common mechanism of adsorption is operating. The results are also suggestive that the value of the distribution coefficient is independent of the chemical composition of the sediment particles, pointing to geochemical control by a surface property rather than one of the bulk sediment. This agrees with the evidence from the sediment extraction experiment where plutonium was found to be associated with the reducible oxide coatings or layers.

Recent experimental work using a double tracer experiment (^{238}Pu and ^{239,240}Pu) has shown that the reaction leading to the measured distribution coefficients in the sediments is reversible [42]. It has also been shown that at Enewetak Atoll the sediments are acting as a source of plutonium in that plutonium is desorbed from the sediments each time the water is exchanged in the lagoon and thereby transported into the Pacific Ocean where it can again be adsorbed by particles and transferred to a different sink [43].

The experimental evidence therefore supports the hypothesis that the concentration of trace elements such as plutonium in the ocean is controlled by an equilibrium distribution process. The values of K_d for the different transuranic elements are summarized in Table V. It is clear that the K_d's vary in the order:

$$Th(IV) \cong Pu(IV) \cong Am(III) > U(VI) \gg Np(V)$$

The value for Pu(IV) is calculated on the basis of the concentration of Pu(IV) in the water column and $K_d[Pu(V)]$ is assumed to be equal to $K_d[Np(V)]$, i.e. effectively no Pu(V) is adsorbed onto sediments [44]. If any changes occur in the ocean which might affect either the redox equilibrium or the concentration of complexing ligands, e.g. in local bays or harbours, then these values of K_d may also change considerably. However, in anoxic interstitial water from the Irish Sea, the value of K_d for Pu(IV) is confirmed since no Pu(V) is measured in these samples [29].

Since plutonium has been shown to exist in fresh and marine waters in a form which adsorbs strongly to anion exchange resins, it is reasonable to suggest that the mechanism of particle uptake involves the adsorption of anionic complexes

TABLE V. COMPARISON OF THE VALUES OF THE DISTRIBUTION
COEFFICIENTS (K_d) FOR LANTHANUM AND ACTINIDE ELEMENTS[a]

Element	La(III)	Th(IV)	U(VI)	Np(V)	Pu(total)	Pu(IV)	Am(III)
Log K_d	5.2	$\geqslant 6.5$	~4.0	$\leqslant 5 \times 10^2$	5.5	>6.5	6.0

[a] See Refs [16] and [44].

of plutonium (or americium) onto iron oxide surface layers. Detailed analyses of
freshwater and marine ferromanganese nodules indicate that anionic and cationic
species are strongly associated with iron oxide and manganese oxide layers
respectively. As there is experimental evidence showing that plutonium (IV) or (VI)
can form strong hydroxyl or carbonato complexes, the stability constants for
other ligands such as SO_4^{-2} and Cl^- are low, and $[CO_3^{-2}]_{total}$ is relatively large in
the oceans or Great Lakes waters, it would seem reasonable to suggest that the
major species involved are carbonato complexes. However, the evidence of higher
concentrations of Pu(III or IV) in less oligotrophic waters and the increasing
tendency for this plutonium to be associated with natural colloids suggests that
this mechanism cannot explain all of the experimental observations.

7.   CONCLUSIONS

    The question of the determination of speciation of the transuranic elements
in natural waters is complex. While a great deal has been learned regarding this
behaviour at extremely low concentrations using very clever and simple experiments, our understanding of the processes involved is limited by our lack of
knowledge of the required fundamental thermodynamic constants for the redox
and equilibrium reactions. Conventional electrochemical and spectrophotometric
techniques cannot be used to determine these properties at the concentration
ranges likely to be encountered in the environment. Because of fundamental
changes in the chemical properties of these elements in terms of the probability
of dimer formation or disproportionation reactions, at very low concentrations,
extrapolation from behaviour at higher concentrations where such reactions
occur will not necessarily provide the correct answers.

    There is no doubt that the behaviour of plutonium in the environment is
strongly affected by the formation of inorganic or organic complexes. At present
very few stability constants have been measured that are applicable to environmental conditions. Techniques must be developed to measure these important
constants. Furthermore, more experiments are needed to understand the
parameters such as the concentration of dissolved oxygen or dissolved organic

carbon which may control the oxidation potential in natural waters and, therefore, the oxidation state of plutonium.

REFERENCES

[1] WAHLGREN, M.A., ROBBINS, J.A., EDGINGTON, D.N., "Plutonium in the Great Lakes", Transuranic Elements in the Environment (HANSON, W., WATTERS, R.A., Eds), US Dept of Energy (1980) 659.

[2] CLEVELAND, J.M., The Chemistry of Plutonium, Gordon and Breach, New York (1970).

[3] JOSEPH, A.B., GUSTAFSSON, P.F., RUSSELL, I.R., SCHUBERT, E.A., VOLCHOK, H.L., TAMPLIN, A., "Sources of radioactivity and their characteristics", Radioactivity in the Marine Environment (1971) 6.

[4] ARKROG, A., Environmental behavior of plutonium accidentally released at Thule, Greenland, Health Phys. 32 (1977) 271.

[5] EDGINGTON, D.N., "A review of the persistence of long-lived radionuclides in the marine environment—sediment/water interactions", Impacts of Radionuclide Releases into the Marine Environment (Proc. Symp. Vienna, 1980), IAEA, Vienna (1981) 67.

[6] PILLAI, K.C., MATHEW, E., "Plutonium in the aquatic environment. Its behaviour, distribution and significance", Transuranic Nuclides in the Environment (Proc. Symp. San Francisco, 1975), IAEA, Vienna (1976) 25.

[7] KATZ, J.J., SEABORG, G.T., The Chemistry of the Actinide Elements, John Wiley and Sons, New York (1958).

[8] NELSON, D.M., ORLANDINI, K.A., "Measurement of Pu(V) in natural waters", Radiological and Environmental Research Division Annual Report, ANL-79-65, Part III (1979) 57.

[9] HOLME, E., "Release of ^{237}Np to the environment: Measurements of marine samples contaminated by different sources", Impacts of Radionuclide Releases into the Marine Environment (Proc. Symp. Vienna, 1980), IAEA, Vienna (1981) 155.

[10] NELSON, D.M., personal communication.

[11] WONG, K.M., Radiochemical determination of plutonium in sea water, sediments and marine organisms, Anal. Chim. Acta 56 (1971) 355.

[12] GOLCHERT, N.W., LEDLET, J., Radiochemical determination of plutonium in environmental water samples, Radiochem. Radioanal. Lett. 12 (1972) 215.

[13] HETHERINGTON, J.A., JEFFERIES, D.F., LOVETT, M.B., "Some investigations into the behaviour of plutonium in the marine environment", Impacts of Nuclear Releases into the Aquatic Environment (Proc. Symp. Otaniemi, 1975), IAEA, Vienna (1976) 193.

[14] WONG, K.M., BROWN, G.S., NOSHKIN, V.E., A rapid procedure for plutonium separation in large volumes of fresh and saline water by manganese dioxide coprecipitation, J. Radioanal. Chem. 42 1 (1978) 7.

[15] INTERNATIONAL ATOMIC ENERGY AGENCY, Reference Methods for Marine Radioactivity Studies II, Technical Reports Series No.169, IAEA, Vienna (1975) 239 pp.

[16] WAHLGREN, M.A., ALBERTS, J.J., NELSON, D.M., ORLANDINI, K.A., "A study of the behaviour of transuranics and possible chemical homologues in Lake Michigan water and biota", Transuranium Nuclides in the Environment (Proc. Symp. San Fracisco, 1975), IAEA, Vienna (1976) 9.

[17] BOWEN, V.T., WONG, K.M., NOSHKIN, V.E., Plutonium-239 in and over the Atlantic Ocean, J. Mar. Res. 29 (1971) 1.

[18] ALBERTS, S.S., WAHLGREN, M.A., NELSON, D.M., JEHN, P. J., Submicron particle size and change characteristics of ^{239,240}Pu in natural waters, Environ. Sci. Technol. **11** (1977) 673.

[19] NOSHKIN, V.E., WONG, K.M., EAGLE, R.J., GATROUSIS, C., Transuranics and other radionuclides in Bikini Lagoon: concentration data retrieved from aged coral sections, Limnol. Oceanogr. **20** (1975) 729.

[20] WATTERS, R.A., EDGINGTON, D.N., HAKONSON, T.E., HANSON, W.C., SMITH, M.H., WHICKER, F.W., WILDUNG, R.E., "Synthesis of the research literature", Transuranic Elements in the Environment (HANSON, W., WATTERS, R.A., Eds), US Dept of Energy (1980) 1.

[21] POLZER, W.L., "Solubility of plutonium in soil/water environments", Safety in Plutonium Handling Facilities (Proc. Symp. Rocky Flats), U.S. AEC CONF-710401 (1971) 411.

[22] SILVER, G.L., "Plutonium in natural waters", Safety in Plutonium Handling Facilities, (Proc. Symp. Rocky Flats), U.S. AEC CONF-710401 (1971) 2.

[23] POURBAIX, M., Atlas of Electrochemical Equilibria, Pergamon Press, New York (1966) 198.

[24] HUTCHINSON, G.E., A Treatise on Limnology, Vol.1, John Wiley and Sons, New York (1957) 691.

[25] LATIMER, W.M., Oxidation Potentials, 2nd Ed., Prentice Hall, New York (1952).

[26] PENNEMAN, R.A., COLEMAN, J.S., KEENAN, T.K., Alkaline oxidation of americium: preparation and reactions of Am(IV) hydroxide, J. Inorg. Nucl. Chem. **17** (1961) 138.

[27] RAI, D., SERNE, R.J., Plutonium activities in soil solutions and the stability and formation of selected minerals, J. Environ. Qual. **6** (1977) 1.

[28] NELSON, D.M., LOVETT, M.B., Oxidation state of plutonium in the Irish Sea, Nature **276** (1978) 599.

[29] NELSON, D.M., LOVETT, M.B., "Measurements of the oxidation state and concentration of plutonium in interstitial waters of the Irish Sea", Impacts of Radionuclide Releases into the Marine Environment (Proc. Symp. Vienna, 1980), IAEA, Vienna (1981) 105.

[30] WAHLGREN, M.A., ALBERTS, J.J., NELSON, D.M., ORLANDINI, K.A., KUCERA, E.T., "Study of the occurrence of multiple oxidation states of plutonium in natural water systems", Radiological and Environmental Research Division Annual Report, ANL-77-65, Part III (1977) 95.

[31] INOUE, Y., TOCHIYAMA, O., Determination of the oxidation states of neptunium at tracer concentrations by adsorption on silica gel and barium sulfate, J. Inorg. Nucl. Chem. **39** (1977) 1443.

[32] BONDIETTI, E.A., TRABALKA, J.K., Evidence for plutonium(V) in an alkaline fresh-water pond, Radiochem. Radioanal. Chem. Lett. **42** (1980) 169.

[33] EDGINGTON, D.N., ROBBINS, J.A., "The behaviour of plutonium and other long-lived radionuclides in Lake Michigan. II. Patterns of deposition in the sediments", Impacts of Nuclear Releases into the Aquatic Environment (Proc. Symp. Otaniemi, 1975), IAEA, Vienna (1975) 245.

[34] EDGINGTON, D.N., ROBBINS, J.A., "Patterns of deposition of natural and fallout radionuclides in the sediments of Lake Michigan and their relation to limnological processes", Environmental Biogeochemistry, Vol.2, Metals Transfer and Ecological Mass Balances (NRIAGU, J.O., Ed.), Ann Arbor Science, Ann Arbor, MI (1976) 705.

[35] ALBERTS, J.J., WAHLGREN, M.A., MULLER, R.N., ORLANDINI, K.A., "The association of transuranic elements with lacustrine sediments", Radiological and Environmental Research Division Annual Report, ANL-75-60, Part III (1975) 36.

[36] EDGINGTON, D.N., ALBERTS, J.J., WAHLGREN, M.A., KARTTUNEN, J.O.,
     REEVE, C.A., "Plutonium and americium in Lake Michigan sediments", Transuranium
     Nuclides in the Environment (Proc. Symp. San Fracisco, 1975), IAEA, Vienna (1976) 493.
[37] CHESTER, R., these Proceedings.
[38] EDGINGTON, D.N., unpublished data.
[39] NOSHKIN, V.E., "Transuranium radionuclides in components of the benthic environ-
     ment of Enewetak Atoll", Transuranic Elements in the Environment (HANSON, W.,
     WATTERS, R.A., Eds), US Dept of Energy (1980) 578.
[40] SCHELL, W.R., SIBLEY, T.H., NEVISSI, A., SANCHEZ, A., Distribution Coefficients for
     Radionuclides in Aquatic Environments. II. Studies on Marine and Freshwater Sediments
     Systems Including the Radionuclides ^{106}Ru, ^{137}Cs, and ^{241}Am, Annual Report August 1977—
     July 1978, U.S. Nucl. Reg. Comm., NUREG/CR-0802 (1979).
[41] DUURSMA, E.K., PARSI, P., "Distribution coefficient of plutonium between sediment
     and sea water", Activities of the International Laboratory of Marine Radioactivity.
     1974 Report, IAEA-163), unpriced document.
[42] EDGINGTON, D.N., KARTTUNEN, J.O., NELSON, D.M., LARSEN, R.P., "Plutonium
     concentration in natural waters — its relationship to sediment adsorption and desorption",
     Radiological and Environmental Research Division Annual Report, ANL-79-65, Part III
     (1979) 54.
[43] NOSHKIN, V.E., WONG, K.M., "Plutonium mobilization from sedimentary sources to
     solution in the marine environment" (Proc. 3rd Nuclear Energy Agency, NEA, Seminar
     in Radio-Ecology, Tokyo, 1979 (in press).
[44] HARVEY, B.R., Potential for post-depositional migration of neptunium in Irish Sea
     sediments", Impacts of Radionuclide Releases into the Marine Environment (Proc.
     Symp. Vienna, 1980), IAEA, Vienna (1981) 93.

DETERMINATION OF SOME OXIDATION STATES OF PLUTONIUM IN SEA WATER AND ASSOCIATED PARTICULATE MATTER

M.B. LOVETT
Ministry of Agriculture, Fisheries and Food,
Directorate of Fisheries Research,
Fisheries Radiobiological Laboratory,
Lowestoft, Suffolk,
United Kingdom

D.M. NELSON
Argonne National Laboratory,
Argonne, Illinois,
United States of America

Abstract

DETERMINATION OF SOME OXIDATION STATES OF PLUTONIUM IN SEA WATER AND ASSOCIATED PARTICULATE MATTER.

A full description is given of an analytical procedure to separate the lower oxidation states of Pu (III and IV) from the higher oxidation states (V and VI). This is possible by the simultaneous use of two isotopic tracers, ^{242}Pu(IV) and ^{236}Pu(VI), which are added as early as possible in the analytical scheme. Various aspects of the chemical processes are discussed, as are the procedures adopted for the preparation of the tracer solutions.

1. INTRODUCTION

The potential importance of the oxidation state in determining the environmental chemistry of plutonium has previously been stated [1–4]. Analytical procedures to separate the oxidation states of plutonium, applied to laboratory samples, have been published [5, 6] but, to our knowledge, no information has been published regarding the oxidation state of plutonium in sea water prior to our investigations [7].

It has been observed at the Fisheries Radiobiological Laboratory, Lowestoft (FRL), that plutonium in the Irish Sea, resulting from the British Nuclear Fuels Ltd. Windscale discharge, is less efficiently removed from sea water onto suspended particulate matter than is americium, which it might well be expected to resemble chemically, the distribution coefficient (K_d = pCi·kg^{-1} particulate matter/pCi·kg^{-1} sea water) for americium being $\sim 10^6$ and for plutonium $\approx 10^5$. Further, it has also been observed that the 'soluble' plutonium (that which passes a 0.22 μm Millipore filter) behaves in a conservative manner with respect to sea water and can be compared to the behaviour of radiocaesium [8].

27

Based upon comparisons with natural chemical analogues, the different oxidation states can be expected to show markedly different biochemical and chemical properties. Thorium and the uranyl ion can be regarded as analogues of Pu(IV) and Pu(VI) respectively and, although the crustal abundance of the elements uranium and thorium is similar, the concentration of uranium in sea water is some 10^4 higher than that of thorium [9]. Similarly, differences in direct biological accumulation are observed, as is stated in the review by Cherry and Shannon [10]. Investigations into the speciation of plutonium in Lake Michigan, USA, conducted at the Argonne National Laboratory (ANL), revealed that the soluble plutonium in Lake Michigan is predominantly (>75%) in one of the higher oxidation states. These factors led to the present joint project between FRL and ANL into the speciation of Pu in United Kingdom coastal waters, with emphasis being placed on the Irish Sea.

2.   DISCUSSION

It is well known that plutonium has several oxidation states in which it can exist in solution, the most common of which are Pu(III), Pu(IV), Pu(V) and Pu(VI). Which of these oxidation states will be found in any particular solution will depend on the nature of that solution, i.e. pH, concentration of plutonium, presence of complexing ions etc. The concentration of plutonium in the sea water around the UK varies from $\approx 10^{-13}$ to $\approx 10^{-16}$M, so the chance of formation of oxidation states by disproportionation reactions becomes remote and therefore the establishment of the oxidation states found can be considered to be the result solely of chemical or biochemical action.

It is accepted that, in general, oxidation state changes involving only electron transfer, e.g. Pu^{3+}(III) $\rightleftharpoons Pu^{4+}$(IV) or PuO_2^+(V) $\rightleftharpoons PuO_2^{2+}$(VI), are faster and more facile than changes that also involve the formation or breaking of the Pu-O linkage. It was considered that, in a well-oxygenated medium such as sea water, the higher oxidation state in the pairs Pu(III), Pu(IV) or Pu(V), Pu(VI) were the more likely, so henceforward the lower oxidation states, Pu(III) and Pu(IV), will be referred to as Pu(IV), and the higher oxidation states, Pu(V) and Pu(VI), will be referred to as Pu(VI).

The chemical basis of the procedure used hinges on three main points:

(a) two isotopic tracers, ^{242}Pu(IV) and ^{236}Pu(VI), are added at the outset;

(b) Pu(IV) will quantitatively coprecipitate on a rare earth fluoride [11] under appropriate conditions, whereas Pu(VI) will not coprecipitate;

(c) a holding oxidant, $K_2Cr_2O_7$, is added to prevent reduction of Pu(VI) by any reducing agent incidentally added for other purposes, e.g. in the HF added to precipitate the rare earth fluoride. This holding oxidant should also oxidize any Pu(V) to Pu(VI) (as well as Pu(III) to Pu(IV)).

TABLE I. OXIDATION RATE OF Pu(IV) BY HOLDING OXIDANT IN
EITHER A SEA WATER OR A DISTILLED WATER MEDIUM

Time	^{242}Pu(IV) tracer in the ^{236}Pu(VI) fraction		^{236}Pu(VI) tracer in the ^{242}Pu(IV) fraction	
	Sea water (%)	Distilled water (%)	Sea water (%)	Distilled water (%)
At zero time	1.0	1.4	1.0	0.7
After 9 days	1.6	9.5	0.7	0.5
After 21 days	2.0	13.5	0.8	1.0

Note: The experiment was carried out by adding the two tracers, ^{242}Pu(IV) and ^{236}Pu(VI) to sea water and distilled water to which the prescribed acids $K_2Cr_2O_7$ and Nd had been added.

3. OUTLINE OF THE METHOD

The two tracers are added as soon as possible after collection or after filtration in the case of filtered samples (Appendix A). After thorough mixing, each sample is made 0.8M in HNO_3, 0.25M in H_2SO_4, 0.0005M in $K_2Cr_2O_7$ and 0.0007M in Nd. The acids are added, firstly, to produce suitable conditions for the quantitative coprecipitation of Pu(IV) and non-coprecipitation of Pu(VI), secondly, to break up any hydrolysis products or complexes of plutonium in the sea water and, thirdly, to solubilize the exchangeable plutonium associated with particulate matter in the unfiltered samples. The dichromate is also capable of oxidizing Pu(IV) to Pu(VI), but the process is slow in sea water ($\approx 0.25\%$ per week, although faster, up to 1% per day, in a freshwater medium — see Table I) and can be ignored because in the worst case, that of unfiltered sea water, samples are not allowed to stand at this stage for more than ≈ 16 hours (overnight).

Unfiltered samples are allowed to stand overnight to allow the exchangeable plutonium to leach from the particulate matter, which is then removed by filtration. An alternative procedure is to filter the sample and then to leach the particles immediately by recirculating through the filter about 4 ltr of a solution containing the tracers and acids, neodymium and dichromate in the prescribed amounts. This leach is then treated as a filtered water sample.

The sample is now made 0.25M in HF and, after allowing the NdF_3 precipitate to develop (≈ 15 minutes usually), it is removed by filtration and reserved for subsequent Pu(IV) determination.

TABLE II. REDUCTION RATE OF
^{239}Pu(VI) BY THE FERROUS ION
($\equiv$2 g FERROUS AMMONIUM
SULPHATE/LITRE)

Time (h)	^{239}Pu on NdF$_3$ after reduction with Fe^{++} (counts/min)
0	354
$\frac{1}{2}$	378
1	375
2	371
4	373

Note: Initial addition: 378 counts/min.

Two grams of ferrous ammonium sulphate are added per litre of filtrate to reduce the Pu(VI). This reduction is fast (see Table II) and is virtually complete in half an hour. After about an hour the reduced solution is made 0.0007M in neodymium and, after standing for $\approx$15 minutes, the precipitated NdF$_3$ is filtered off and reserved for Pu(VI) determination.

Each NdF$_3$ precipitate is treated to remove the fluoride ion: then the plutonium is purified by routine ion-exchange procedures [12]. After electro-deposition, again a standard procedure, onto stainless steel discs, individual plutonium isotopes are determined by α-spectrometry using silicon surface barrier detectors.

The step-by-step procedure for analysis is given in Appendix B.

4.　RESULTS

The advantages and disadvantages associated with either leaching technique can be seen in Table III which shows a fairly typical distribution of the plutonium in an Irish Sea water sample. 80% of the plutonium is removed from the particulate material, and the fact that some sort of equilibrium has been established is illustrated in Table IV. We do not find any ^{236}Pu(VI) tracer on the leached particulates and infer from this the complete removal of the hexavalent phase. The median values and ranges for chemical recoveries and degree of cross-contamination for some of the analyses we have performed and an indication of sample size are shown in Table V.

TABLE III. DISTRIBUTION OF PLUTONIUM SPECIES IN A SAMPLE OF IRISH SEA WATER (%)

Pu species	A Whole sea water	B Filtrate	A−B Particulate matter
Pu(IV)	74.2	1.8	72.4
Pu(VI)	25.8	25.1	0.7

TABLE IV. ESTABLISHMENT OF ISOTOPIC EQUILIBRIA

	$\dfrac{^{239}Pu\ (sample)}{^{242}Pu\ (tracer)}$	
	Remaining on particulate matter	Leach
Whole sea water method	11	7
	5	3
	2	3
Leaching filtered particles	32	27
	23	44
	3.8	1.4

TABLE V. RESULTS OBTAINED FROM ENVIRONMENTAL SAMPLES: MEDIAN VALUES (AND RANGES) OF A NUMBER OF DETERMINATIONS

Sample volume	Pu(IV) fraction		Pu(VI) fraction	
	Recovery (%)	$^{236}Pu(VI)$ tracer present (%)	Recovery (%)	$^{242}Pu(IV)$ tracer present (%)
20−50 ltr	91 (34−100)	1.2 (0−9.9)	87 (48−100)	2.2 (0.5−2.2)
1 ltr	87 (73−96)	0.6 (0.2−0.9)	83 (24−100)	1.0 (0.5−2.9)
100 ml	90 (23−100)	0.4 (0.2−0.8)	82 (16−100)	1.6 (0.2−3.2)

5.   CONCLUSIONS

The main conclusions to be drawn are:

(a) The separations of the two oxidation states is quite adequate and in general the degree of cross-contamination is small. The use of the two tracers means that correction for the indigenous sample plutonium in the 'wrong' fraction can be made.

(b) The overall chemical recoveries are very good, the occasional poor recovery can usually be traced back to the electrodeposition stage.

REFERENCES

[1]   ANDELMAN, J.B., ROZZELL, T.C., Radionuclides in the environment, Adv. Chem. Ser. **93** (1970).
[2]   NOSHKIN, V.E., Health Phys. **22** (1972) 537.
[3]   SCHELL, W.R., WATTERS, R.L., Health Phys. **29** (1975) 589.
[4]   DAHLMAN, R.C., BONDIETTI, E.A., EYMANN, L.D., Actinides in the Environment, A. Chem. Soc. Symp. Ser. **35** (1976) 47.
[5]   FOTI, S.C., FREILING, E.C., Talanta **11** (1964) 385.
[6]   BONDIETTI, E.A., REYNOLDS, S.A., Proceedings of an Actinide-Sediment Reactions Working Meeting at Seattle, Washington, 10–11 Feb. 1976, BNWL 2117 (1976) 505.
[7]   NELSON, D.M., LOVETT, M.B., Nature **5688** (1978) 599.
[8]   HETHERINGTON, J.A., JEFFERIES, D.F., LOVETT, M.B., "Some investigations into the behaviour of plutonium in the marine environment", Impacts of Nuclear Releases into the Aquatic Environment (Proc. Symp. Otaniemi, 1975), IAEA, Vienna (1975) 193.
[9]   BURTON, J.D., Chemical Oceanography (RILEY, J.P., SKIRROW, G., Eds) Vol.3, Academic Press, London (1975) 97.
[10] CHERRY, R.D., SHANNON, L.V., At. Energy Rev. **12** (1975) 3.
[11] MAGNUSSON, L.B., LA CHAPELLE, T.J., J. Am. Chem. Soc. **70** (1948) 3534.
[12] NELSON, D.M., YAGUCHI, E.M., WALLER, B.J., WAHLGREN, M.A., Radiological and Environmental Research Division Annual Report, 1973, Argonne National Laboratory, Argonne, Ill., Rep. ANL-8060, part III (1973) 6.

Appendix A

TRACER SOLUTIONS

Tracer solutions prepared as described here should be at least 99% pure and have been found to be stable for at least a year.

Preparation of ^{242}Pu(IV) tracer solution

Add ^{242}Pu to 16M HNO$_3$ and evaporate to dryness. Dissolve in 8M HNO$_3$ and pass through an anion exchange column (Bio-rad AG1X4, 100–200 mesh, NO$_3^+$ form, 50 mm X 7 mm diameter). Wash well with 8M HNO$_3$ and discard washings. Elute ^{242}Pu(IV) with 10 ml 0.1M HNO$_3$ and immediately add 10 ml 16M HNO$_3$ to the eluate. Dilute with 8M HNO$_3$ to the required volume.

Preparation of ^{236}Pu(VI) tracer solution

Evaporate ^{236}Pu solution to dryness. Dissolve in 2–3 ml 0.8M HNO$_3$ and transfer to the anode compartment of the electrolysis cell (Fig.A-1). Fill the cathode compartment of the cell to the same depth as the anode compartment with 0.8M HNO$_3$. Pass direct current at 3.5 V overnight (16 h). Disconnect, remove the anode solution and dilute to the required volume with 0.8M HNO$_3$.

It should be noted that approximately one third of the plutonium originally added to the anode compartment will be lost from this compartment to the cathode compartment during the electrolysis. Allowance for this should be made when deciding on the amount of tracer ultimately required in the Pu(VI) oxidation state. Any Pu in the cathode compartment, mostly in the reduced state, can, of course, be recovered.

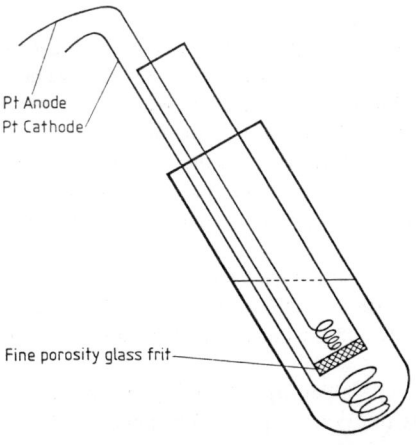

Pt Anode
Pt Cathode

Fine porosity glass frit

FIG.A-1.   Apparatus used in the preparation of Pu(VI) tracer.

Appendix B

DETAILS OF ANALYTICAL METHOD

The use of distilled or demineralized water is implied throughout.

Special reagents required

(1) ^{242}Pu(IV) tracer
(2) ^{236}Pu(VI) tracer } see Appendix A for details of preparation.
(3)   Holding oxidant solution — dissolve 2.94 g $K_2Cr_2O_7$ per litre 5M H_2SO_4.
(4)   Nd solution — dissolve 117 g Nd_2O_3 per litre 8M HNO_3 to give a solution which is 0.7M in Nd (100 mg Nd/ml).
(5)   Al solution — a saturated solution of $Al(NO_3)_3$ in 8M HNO_3.
(6)   Anion exchange resin — Bio-rad AG1 X 4, 100–200 mesh.
(7)   Electrodeposition solution — 1.26 g ammonium oxalate and 53.5 g ammonium chloride dissolved in 1 ltr water.

The quantities of reagents used in the method are those used for a sample size of 45 ltr which is the largest volume analysed to date.

Method

(1)   To the sample (see note 1 below) add appropriate quantities of ^{242}Pu(IV) and ^{236}Pu(VI) tracers, 2.5 ltr 16M HNO_3, 2.5 ltr holding oxidant solution and 50 ml Nd solution, mixing well after each addition. Allow to stand for about 30 min (see note 2 below).

(2)   Add 500 ml 40% wt/vol. HF, mix well and stand for 15 min. Filter and reserve the precipitate for subsequent purification and estimation of Pu(IV) in the sample (see steps 4 et seq).

(3)   To the filtrate add 100 g ferrous ammonium sulphate, mix well and stand for 30 min. Add 50 ml Nd solution, mix well, stand for ≈15 min and filter. Discard the filtrate. Reserve the precipitate for subsequent purification and estimation of Pu(VI) in the sample.

(4)   Transfer the filter paper and NdF_3 to a beaker containing 2 g H_3BO_3 and ≈70 ml 16M HNO_3. Cover and boil gently until the filter paper has been completely destroyed and the precipitate solubilized (see note 3 below).

(5)   Dilute with water to ≈140 ml and warm to dissolve any crystalline deposits and transfer to a 1 ltr centrifuge bottle. Add, with stirring, 90 ml 0.880 ammonia solution which has been diluted to ≈500 ml. Centrifuge and discard supernate.

(6)   Dissolve the precipitate in sufficient 16M HNO_3 to give a solution ≈8M in HNO_3. Add 1 ml Al solution.

(7)   Pass the solution through an anion exchange resin column (50 mm X 7 mm diameter) previously equilibrated with 8M HNO_3. Wash with 25 ml 8M HNO_3 then 25 ml 11M HCl. Discard the raffinate and washings. Elute the plutonium with ≈6 ml 11M HCl/0.1M NH_4I.

(8)   Evaporate to dryness, add ≈20 drops 16M HNO_3 and evaporate to dryness. Add 1 ml 11M HCl and evaporate to dryness. To the warm beaker add 5 drops 3M HCl to dissolve the plutonium and transfer to an electrodeposition cell (see note 4 below) using 10 ml electrodeposition solution. Pass a current of 800 mA for 45 min. Before

disconnecting the current add 1 ml 0.880 ammonia solution and wait 1 min. Remove
the stainless steel disc, rinse carefully with water then methylated spirit and heat
in a flame.

Notes

(1) The sample is either filtered or unfiltered sea water.
(2) If unfiltered, it is stood for ≈ 16 h before any undissolved particulate matter is
filtered off.
(3) If only a small sample, e.g. 1 ltr or less, is being analysed, it is sufficient to destroy
the paper by boiling to incipient dryness with 16M HNO_3. Dissolve, heating if
necessary, in 10 ml 8M HNO_3 and 1 ml Al solution.
(4) The cell has a 19 mm diameter stainless steel disc as a cathode, the area of deposition
is ≈ 200 mm^2. Pt is used as an anode. Since Cl_2 gas is evolved, this electrodeposition
should be carried out under a fume hood.

SPECIATION OF PLUTONIUM IN THE MEDITERRANEAN ENVIRONMENT

R. FUKAI, A. YAMATO*, M. THEIN
International Laboratory of Marine Radioactivity, IAEA,
Musée Océanographique,
Principality of Monaco

H. BILINSKI
Ruder Bošković Institute,
Zagreb,
Yugoslavia

Abstract

SPECIATION OF PLUTONIUM IN THE MEDITERRANEAN ENVIRONMENT.

Based upon thermodynamical computations, plutonium in well-oxygenated aqueous solution at pH 5.5 is estimated to be ~80% as PuO_2^+ (Pu(V)) and ~20% as PuO_2^{2+} (Pu(VI)), ignoring complex formation. The results of the measurements of the oxidation states of plutonium in rain water collected at Monaco during 1979 agree reasonably well with the above values, assuming the disproportionation of Pu(V) during the storage of the rain samples. The different behaviour of plutonium as opposed to americium in the Mediterranean water column is interpreted in the light of the chemical speciation of plutonium in the rain input into the Mediterranean. This qualitatively explains why plutonium is more stable than americium in the Mediterranean water column, but does not cover entirely the quantitative difference in vertical transport between these two actinides. Possible stabilization of Pu(IV) by the formation of its soluble complexes is suggested.

1. INTRODUCTION

The measurements of the oxidation states of fallout plutonium contained in rain collected at Monaco during 1979 have shown that a substantial fraction of plutonium occurs in the higher oxidation states (Pu(V) and Pu(VI)), although the percentage of the higher valency plutonium tends to decrease with increasing lengths of storage between rainfall and the chemical treatment [1]. In order to understand the speciation dynamics of plutonium in rain, computations on the thermodynamical equilibria of various oxidation states in an aqueous solution have been carried out. In the present paper the basis and outline of these computations are given. Based on these computations the significance of the occurrence of the higher-valency plutonium in sea water is discussed in relation

---

* Present Address: Power Reactor and Nuclear Fuel Development Corporation, Tokai Works, Tokai-mura, Japan.

to the characteristic stability of fallout plutonium in the Mediterranean water column.

2.   OUTLINE OF COMPUTATIONS

Since the major source of plutonium present today in the Mediterranean Sea is radioactive fallout injected into the atmosphere by nuclear detonation tests, and also since more than 90% of the fallout delivery of plutonium is brought down by rain [2], the chemical speciation of plutonium in rain water should influence substantially the behaviour of this actinide in Mediterranean sea water.

Oxido-reduction of any ions in an aqueous solution should be dependent only on the electron activity and hydrogen ion concentration of the medium, if other types of reactions, such as complex formation, heterogeneous surface reactions, etc., can be ignored. Since the ionic strength of rain water is much lower than that of other types of natural waters, theoretical computations based on thermodynamical constants are considered to fit better for rain water than for other types of natural waters. To estimate the electron activity of an aqueous solution the following equation can be used:

$$\tfrac{1}{2}O_2(g) + 2H^+ + 2e^- = H_2O(1) \qquad \log K_1 = 41.55 \qquad (1)$$

As rain water is well oxygenated, the partial pressure of oxygen in rain is considered to be equal to that in the atmosphere, $pO_2 = 0.21$ atm. The pH values measured for Monaco rain were always around 5.5, showing little variation. Thus, the electron activity $pE = -\log e^-$ is calculated to be 15.11 by using Eq.(1).

Ignoring complex formation, the oxido-reduction equilibria of various oxidation states of plutonium in an aqueous solution can be expressed as follows, taking into account the stability constants compiled by Sillén and Martell [3]:

$$PuO_2^{2+} + e^- = PuO_2^+ \qquad\qquad \log K_2 = 15.7 \qquad (2)$$

$$PuO_2^{2+} + 4H^+ + 2e^- = Pu^{4+} + 2H_2O \qquad \log K_3 = 35.2 \qquad (3)$$

$$Pu^{4+} + e^- = Pu^{3+} \qquad\qquad \log K_4 = 16.38 \qquad (4)$$

The following relationship can be written for the various oxidation states of plutonium in the solution:

$$PuO_2^{2+} + PuO_2^+ + Pu^{4+} + Pu^{3+} = [Pu]_{tot} \qquad (5)$$

where $[Pu]_{tot}$ is the total concentration of soluble plutonium. Since $[e^-]$, $[H^+]$ and $[Pu]_{tot}$ are known, Eqs (2–5) form a system of four equations containing four unknowns, $[PuO_2^{2+}]$, $[PuO_2^+]$, $[Pu^{4+}]$ and $[Pu^{3+}]$. By solving this system of equations one obtains as the fractions of the soluble plutonium 79.55% PuO_2^+ (Pu(V)) and 20.45% PuO_2^{2+} (Pu(VI)) with negligible contributions of Pu^{4+} and Pu^{3+}. These equilibria are pH-sensitive; if the rain water is more acidic, e.g. pH 5.0, then the partition would be 72.46% PuO_2^{2+} (Pu(VI)) and 27.53% PuO_2^+ (Pu(V)). In any case, the above computations show that, in Monaco rain, fallout plutonium is predicted to be ~80% Pu(V) and ~20% Pu(VI) when the equilibria are reached. This indicates that, if the system is at equilibrium, only the presence of the higher valency plutonium (Pu(V) + Pu(VI)) would be detected, when the method of differential measurements of the plutonium oxidation states [4] is applied to Monaco rain water. The method differentiates only between the lower valency plutonium (Pu(III) + Pu(IV)) and the higher valency plutonium (Pu(V) + Pu(VI)).

3. DISCUSSION

In Fig.1 the percentages of the higher valency plutonium measured for Monaco rain during 1979 [1] were plotted against the lengths of the storage of the rain samples before the chemical treatment. The curve in the figure was drawn by assuming that, at the beginning, the percentage was at 100%. Although the points in the figure scatter irregularly, a general trend of the decrease of the percentage with increasing storage time is clearly visible. This suggests that reduction of the higher valency plutonium took place during the storage of rain samples after the arrival of rain at the earth's surface, despite the fact that initial partition of plutonium had been mostly in the higher oxidation states.

Possible mechanisms for the reduction of the higher valency plutonium to the lower oxidation states during the storage of the rain samples cannot be discerned from the data available. As the rain samples collected contained possible reductants such as pollen, organic dusts, soil particles, etc., the reduction of the higher valency plutonium by these reductants may have occurred during the storage. However, Pu(V) is known to disproportionate relatively easily at pH below 1.5 or above 5 [5], according to the following equation:

$$PuO_2^+ + PuO_2^+ + 4H^+ = Pu^{4+} + Pu_2^{2+} + 2H_2O$$

Thus, it is also possible that the disproportionation of Pu(V) would account for the reduction, if the disproportionation reaction could proceed with reasonable speed in an extremely diluted solution of 10^{-15} M or less, like rain water. Although information on the kinetics of the disproportionation reaction in an extremely diluted solution is not available, the very low ionic strength of rain

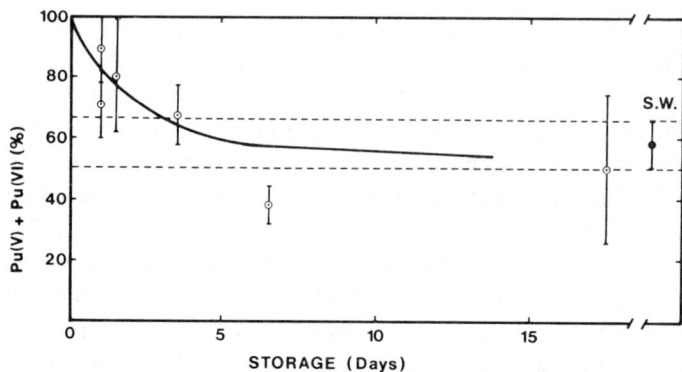

FIG.1. Decrease of the high-valency plutonium fraction (Pu(V) + Pu(VI)) with storage time. S.W. (●) and broken lines indicate respectively the average value for sea water and its range (1σ).

water is considered to favour such a reaction. As shown in the above equation, the disproportionation of Pu(V) results in equal quantities of Pu(IV) and Pu(VI). This indicates that the rain water samples would contain theoretically ~40% Pu(IV) and 60% Pu(VI), if complete disproportionation of Pu(V) took place during storage. In any case, the substantial occurrence of the higher valency plutonium in Monaco rain as well as the decreasing trend of higher oxidation states during storage suggest that the major fraction of fallout plutonium in rain arriving at the Mediterranean surface is in the higher oxidation states, although the equilibria of its partition in sea water afterwards depends on various factors such as kinetics of complex formation, oxido-reduction, disproportionation, etc. It is noted, however, that the decrease of the percentage of the higher valency plutonium with the storage time tends to level off within a percentage range for Mediterranean sea water, as shown in Fig.1. The reasonable agreement of the range for sea water with the value for the higher valency plutonium (~60% Pu(VI)) predicted theoretically for aged rain seems significant, although the range given for sea water is based on a small number of measurements.

The characteristic difference in the vertical transport behaviour between plutonium and americium in the Mediterranean water column has already been well demonstrated [6–8]. The measurements of the oxidation states of plutonium in Mediterranean rain and sea water as well as the thermodynamical computations of speciation equilibria presented above indicate that the different behaviour of plutonium and americium may be attributed to the difference in in situ oxidation states of these two actinides. While it is probable that a major fraction of plutonium in the rain delivery and surface sea water occurs in the higher oxidation states, the chemistry of americium excludes the occurrence of this actinide in the higher oxidation states under natural conditions [9]. Since Pu(VI), like U(VI), is known to form a very stable complex with carbonate, it would remain in solution in the sea water medium for a fairly long time. On the other hand, americium in the lower oxidation states would easily be hydrolysed, associated with negatively charged sinking particles and transported downwards with these particles. The fate of Pu(IV) in sea water is considered to be similar to that of americium. However, the substantial difference in the vertical transport behaviour observed between plutonium and americium in the Mediterranean column does not seem to be entirely explained by taking into account the partial presence of 50 to 60% of soluble Pu(VI) in Mediterranean surface water. A possible stabilization of Pu(IV) by some complex formation cannot be excluded. Further studies of possible complexation of plutonium in the sea water medium will be presented elsewhere.

REFERENCES

[1]   FUKAI, R., YAMATO, A., THEIN, M., BILINSKI, H., Oxidation state of fallout plutonium in Mediterranean rain and seawater (in preparation).

[2]   THEIN, M., BALLESTRA, S., YAMATO, A., FUKAI, R., Delivery of transuranic elements by rain to the Mediterranean Sea, Geochim. Cosmochim. Acta 44 (1980).

[3]   SILLEN, L.G., MARTELL, A.E., Stability Constants of Metal-Ion Complexes, Spec. Publ. No.17, Chem. Soc., London (1964) 7.

[4]   NELSON, D.M., LOVETT, M.B., Oxidation state of plutonium in the Irish Sea, Nature 276 (1978) 599.

[5]   CLEAVELAND, J.M., The Chemistry of Plutonium, Gordon & Breach Sci. Publ. Inc., New York (1970) 11.

[6]   FUKAI, R., BALLESTRA, S., HOLM, E., [241]Americium in Mediterranean surface waters, Nature 264 (1976) 739.

[7]   FUKAI, R., HOLM, E., BALLESTRA, S., A note on vertical distribution of plutonium and americium in the Mediterranean Sea, Oceanol. Acta 2 (1979) 129.

[8]   HOLM, E., BALLESTRA, S., FUKAI, R., BEASLEY, T.M., Particulate plutonium and americium in Mediterranean surface waters, Oceanol. Acta 3 (1980) 157.

[9]   SCHULZ, W.W., The Chemistry of Americium, Tech. Inform. Center, ERDA, Oak Ridge, Tenn. (1976) 47.

METHOD FOR THE DETERMINATION OF ^{237}Np IN LOW-LEVEL ENVIRONMENTAL SAMPLES

E. HOLM, M. NILSSON
Radiation Physics Department,
University of Lund,
Lund, Sweden

Abstract

METHOD FOR THE DETERMINATION OF ^{237}Np IN LOW-LEVEL ENVIRONMENTAL SAMPLES.

A method for the determination of ^{237}Np in environmental samples was developed. A radiochemical procedure based on isolation of neptunium from other interfering α-particle emitters by precipitations and ion exchange was employed. The radiochemical yield was determined by use of ^{239}Np and Ge(Li) spectrometry. ^{237}Np was counted with silicon surface barrier detectors. Biological samples up to 200 g dry weight were processed with a minimal detectable activity of the order of 10 fCi at 10^6 s counting time.

1. INTRODUCTION

Neptunium-237 ($T_{1/2} = 2.1 \times 10^6$ a) is the remaining transuranium element in global fallout from nuclear detonation tests which has not been investigated.

This is certainly due to analytical problems consisting of low activity concentrations in environmental samples and low concentration factors in food chains. To measure very low activities, α-spectrometry must be employed. Peak interferences from other α-particle emitters such as ^{234}U, ^{230}Th, ^{231}Pa and $^{239+240}$Pu present at much higher concentrations in our samples must be considered. By mass, the amount of ^{237}Np present in the environment can be expected to be of the same order as ^{239}Pu since the (n, 2n)/(n, γ) reaction ratio for ^{238}U at a thermonuclear device can be as high as 0.5−1 [1]. At an equal state of oxidation neptunium and plutonium chemically resemble each other but neptunium is frequently present at one oxidation state higher. Being the daughter product of ^{241}Am, ^{237}Np is considered a serious problem in the long-term storage of wastes from nuclear industry.

Several methods have been described for the radiochemical separation of neptunium, anion exchange, cation exchange, liquid/liquid extraction, neutron activation etc. [2]. The method described here was developed in order to investigate ^{237}Np in the environment. Special attention was paid to the problems of yield determination and interfering α-particle emitters.

43

2.   METHOD

2.1. Use of radiochemical yield determinant

As radiochemical yield determinant about 9 Bq (-250 pCi) of ^{239}Np from
a stock solution of ^{243}Am was used. Since plutonium determination is carried
out simultaneously $\sim$180 mBq ($\sim$5 pCi) of ^{242}Pu was used for this purpose as
yield determinant. If americium is in the +3 state and neptunium kept in the +4
state they both precipitate quantitatively at the precipitations described below.
It was thus not necessary to separate the ^{239}Np from the ^{243}Am before adding
to the sample. Neptunium and americium will be separated at the first ion
exchange procedure.

2.2. Separation procedure

The dried samples were ashed at 550°C until no traces of carbon remained.
After leaching with aqua regia and hydrogen peroxide (30%), neptunium (+4)
was precipitated with 20 mg lanthanum as fluoride. The precipitation from the
centrifuged sample was dissolved by heating in nitric acid and a small amount of
boric acid (added as solid). A second precipitation was carried out by adding
ammonia, until precipitation started, then adding saturated ammonium carbonate
solution. The supernate was again discarded and the precipitate dried at 105°C.
Neptunium was kept in the +4 state with ferro-ammonium sulphate during the
procedure. From a solution containing 9M hydrochloric acid and 0.1M ammonium
iodide solution, neptunium was sorbed on an anion exchange resin (8 cm^3;
BIORAD 1 × 8, 100−200 mesh). The effluent and 30 cm^3 9M hydrochloric acid
washing contain plutonium, americium, thorium and the lanthanum added.
Remaining impurities were removed from the column with 30 cm^3 of a solution
containing 1M nitric acid and 93 vol.% methanol. Neptunium was eluted with
4.5M hydrochloric acid and 0.05M hydrogen fluoride. After evaporation with
boric acid and sulphuric acid (1 cm^3) the samples were electrodeposited onto
stainless steel discs from an ammonium sulphate solution as described by Talvitie [3].
For most biological samples a neptunium yield of about 50% can be obtained by
this method. However, the radiochemical yield tended to be much lower when
applying the method to soil and sediment samples. This is probably due to the
presence of aluminium preventing americium and neptunium from precipitating
as fluorides.

The problem was solved by avoiding this fluoride precipitation. After loading
the column neptunium was directly eluted with a solution containing 4.5M
hydrochloric acid and 0.05M hydrogen fluoride. After evaporation with a
solution containing nitric and boric acids, neptunium was extracted into 0.5M TTA
(thenoyltrifluoroacetone) and xylene solution and back-extracted with 8M nitric
acid. The drawback of this procedure is that there is not the same effective

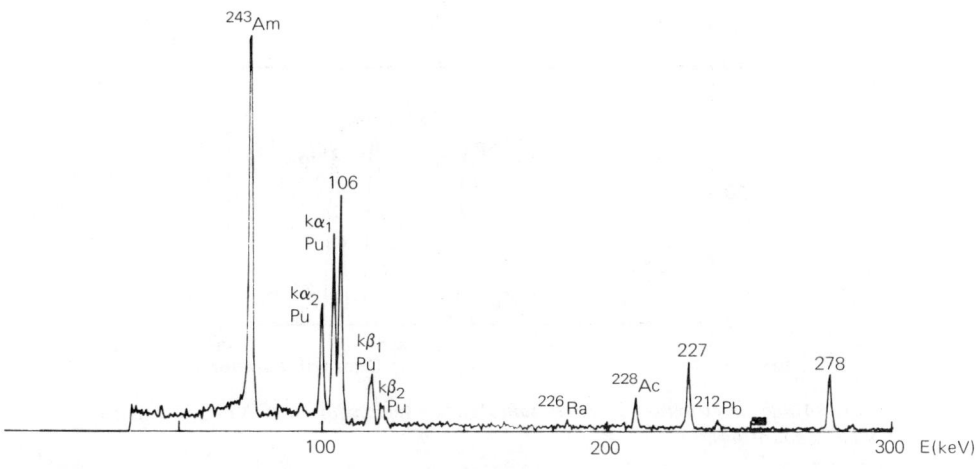

FIG.1. Ge(Li) spectrometry: $^{243}Am-^{239}Np$-*standard.*

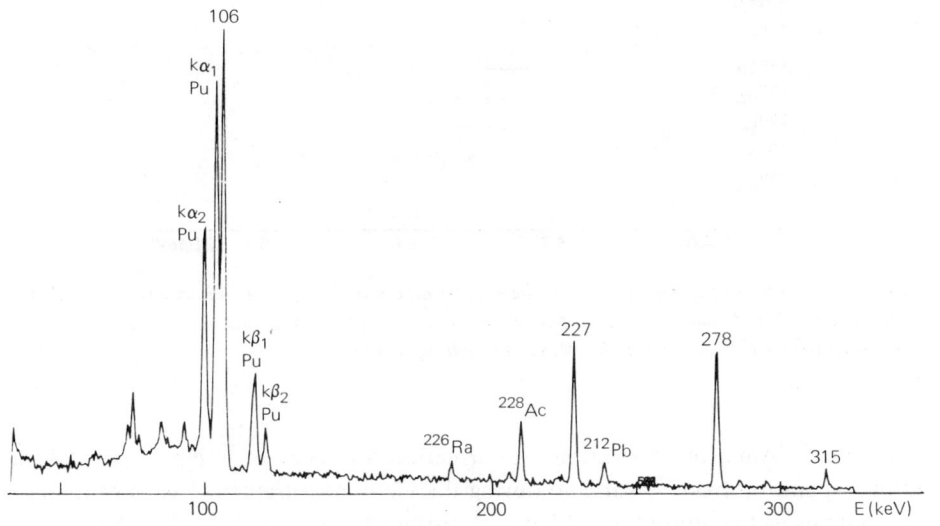

FIG.2. Ge(Li) spectrometry: $^{243}Am-^{239}Np$ *sample after separation.*

reduction of sample sizes as obtained using fluoride precipitation, and larger
volumes must be used in column operation.

2.3. Measurements

The yield was determined by Ge(Li) spectrometry of the sample compared
with a $^{243}Am-^{239}Np$ standard calibrated by alpha spectrometry. The $k_{\alpha 1}$ Pu,
$k_{\alpha 2}$ Pu and 106, 227 and 278 keV from ^{239}Np peaks were integrated. The Ge(Li)
spectra of the standard and of a sample are shown in Figs 1 and 2. Note the

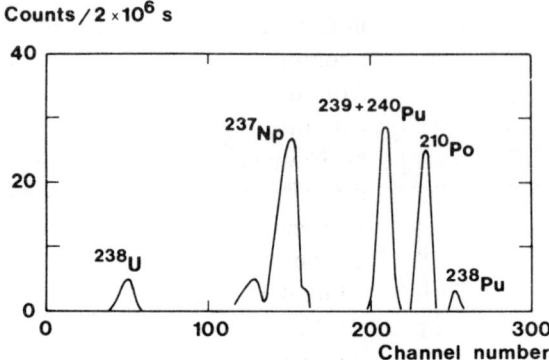

FIG.3. *Alpha spectrogram of lichen sample after separation of* ^{237}*Np (150 g sample,* 2×10^6 *s counting).*

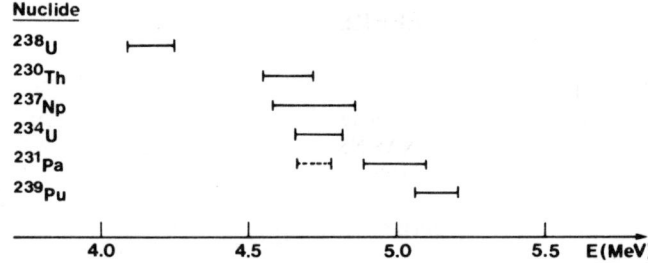

FIG.4. *Actinides of importance at alpha spectrometry of* ^{237}*Np. The energy region which they occupy is indicated. (Surface barrier detector, ORTEC BA-18-300-100, 512 channels per detector, 6 keV per channel.) Minor branching dotted.*

absence of ^{243}Am after separation. A correction for decay of ^{239}Np ($T_{1/2} = 2.35$ d) during separation from ^{243}Am and during the yield measurement has to be made. The samples were counted for ^{237}Np with surface barrier detectors for about 10^6 seconds. A typical α-spectrum is shown in Fig.3.

3.   DISCUSSION

The actinides that must be considered are indicated in Fig.4, as is their interference with ^{237}Np measurements.

The major interfering nuclide is ^{234}U, present in large amounts in the environment and having the same main α-energy as ^{237}Np. The activity ratio ^{234}U/^{237}Np can be as high as 5×10^5 in certain environmental samples. Remaining traces of ^{234}U are corrected for by observing the ^{238}U also present (see Fig.3). The procedure described gives excellent separation from uranium. Uranium does not precipitate as fluoride or carbonate and is not complexed by the hydrogen fluoride as neptunium is at the elution stage from the column. Plutonium-239+240 might interfere owing to tailing of the peak into the ^{237}Np energy region. In the column procedure, plutonium is kept in the 3+ valence state and will not be sorbed. Only a minor non-interfering fraction of $^{239+240}$Pu will appear in the spectra.

Protactinium-231 has two major energies, of which one (12% branching) will interfere in the ^{237}Np energy region. In addition ^{231}Pa could possibly follow neptunium in the separation procedure described. This problem was solved by integrating ^{237}Np over 4.70–4.83 MeV. This will contain 81% of the ^{237}N peak and almost no ^{231}Pa. This energy integration also avoids interference from ^{230}Th, although thorium should not be sorbed together with neptunium during column separation.

REFERENCES

[1] de GEER, L.E., Personal communication, 1979.
[2] BURNEY, G.A., HARBOUR, R.M., The radiochemistry of neptunium Natl. Acad. of Sciences – Natl. Research Council NAS-NS-3060 (1974).
[3] TALVITIE, N.A., Anal. Chem. 44 (1972) 280.

METHODS USED AT CNEN
FOR THE ANALYSIS OF Pu AND [241]Am
IN MARINE SAMPLES

A. DELLE SITE, V. MARCHIONNI
Comitato Nazionale
  per l'Energia Nucleare,
Centro di Studi Nucleari della Casaccia,
Rome,
Italy

Abstract

METHODS USED AT CNEN FOR THE ANALYSIS OF Pu AND [241]Am IN MARINE
SAMPLES.
    Very low levels of plutonium ([239, 240]Pu and [238]Pu) in marine samples (sea water, sediments, marine organisms) are determined by extraction chromatography with tri-n-octyl phosphine oxide (TOPO) supported on microporous polyethylene, electrodeposition and α-spectrometry. [236]Pu or [242]Pu is added as the yield detector and a high-resolution α-spectrometer is used for counting. The final recoveries are 62.6% ± 9.7 (σ) for sea water, 45.4% ± 9.6 (σ) for sediments and 81.7% ± 4.5 (σ) for marine organisms. The method enables [239, 240]Pu and [238]Pu to be detected at the femtocurie level. A method proposed for americium analysis in 50 g sediments is also described. After leaching of the samples, uranium, plutonium and thorium are extracted with TOPO, leaving americium, curium, lanthanides and iron in the aqueous phase. The organic phase is processed for plutonium, while in the aqueous solution separation of transplutonium elements from iron is obtained by a double oxalate co-precipitation. The decontamination of americium from the other transuranics is improved by extraction chromatography with di-2-ethylhexyl phosphoric acid (HDEHP). The final separation from lanthanides is obtained by a double anion-exchange step in thiocyanate media. Americium is electroplated and counted by α-spectrometry. The recovery ranges between 25 and 35%.

1.   INTRODUCTION

    Concentrations of plutonium and americium in marine samples are principally
due to the environmental pollution caused by fallout from nuclear explosions
and are generally at very low level. Environment samples also contain micro traces
of natural α-emitters (uranium, thorium and their decay products) which complicate
the determination of transuranium nuclides [1].
    The methods reported in the literature for the chemical separation of plutonium
are based on ion-exchange resins [1–5] or liquid/liquid extraction with tertiary
amines [6], organophosphorous compounds [7, 8] and ketones [9, 10]. The
analysis of [241]Am in the environment requires the removal of these natural radionuclides which would interfere with the measurement. In addition, it is also

necessary to decontaminate the sample from impurities, such as lanthanides, which are electroplated with americium and cause spectral degradation [1, 7, 11−14].

This paper describes a method for plutonium determination based on double extraction chromatography with a microporous polyethylene supporting tri-n-octylphosphine oxide (TOPO). This technique has been used previously to separate plutonium from human biological samples [15] and is used for the detection of $^{239, 240}$Pu and ^{238}Pu in environmental samples [16]. The method for the determination of americium in sediments is at present under investigation. The procedure involves a decontamination from uranium, thorium and plutonium by TOPO extraction chromatography, followed by two steps of oxalate co-precipitation to eliminate iron interference, and a column separation of americium by extraction chromatography using di-2-ethylhexyl phosphoric acid (HDEHP) to eliminate other transuranium nuclides. The final purification and separation of americium from lanthanides is carried out by a double anion-exchange separation in thiocyanate solution. According to Stary [17] this method of lanthanide/actinide separation is very efficient. It has been used for americium analysis in urine [18] and environmental samples [11].

2.   EXPERIMENTAL PROCEDURE

2.1. Extraction beds

Extraction slurry A: add dropwise, stirring 2 ml of 0.3M TOPO in cyclohexane to 3 g of microporous polyethylene, Microthene-710; add 30 ml of 4M HNO$_3$ and stir with a magnetic stirrer for 30 minutes. This mixture is used for the batch extraction of plutonium.

Extraction slurry B: add dropwise, stirring 2 ml of 0.2M TOPO in cyclo-hexane to 3 g of Microthene; add 30 ml of 4M HNO$_3$ and stir with a magnetic stirrer for 30 minutes. This mixture is transferred to a chromatographic column for the plutonium purification procedure.

Extraction slurry C: add dropwise, stirring 3 ml of 20% HDEHP in toluene to 4.5 g of Microthene; add 30 ml of 0.01M nitric acid and stir with a magnetic stirrer for 30 minutes. This mixture is transferred to a chromatographic column for the americium purification procedure.

The glass chromatographic columns are 30 cm tall and of 1 cm internal diameter.

2.2. Sample pretreatment

Sea water (50−200 ltr), sediments (100 g dry), and organisms (300 g dry) are treated as described by Wong [2]. Then ^{236}Pu or ^{242}Pu (1−2 dis/min) are added as yield monitors. The residues of the wet-ashing are dissolved in 1 ltr 4M HNO_3 and the plutonium valence is adjusted to four with sodium nitrite.

2.3. Plutonium purification

Stage 1: Add the extraction slurry A to the nitric acid solution obtained from the pretreatment of the samples (c. 4M in nitric acid) and stir with a magnetic stirrer for 1 hour. Filter through a Buchner filter and transfer the slurry containing Pu(IV) to an empty chromatographic column, thus obtaining a bed 10 cm high. Wash with 50 ml of 4M HNO_3 and elute with 80 ml of a solution 6M in hydrochloric acid and 0.02M in ammonium iodide at a flow rate of 0.25 ml·min^{-1}. This solution reduces plutonium to Pu(III).

Stage 2: Take the eluate to dryness, add a few drops of concentrated HNO_3 to eliminate iodine completely and dissolve the residue in 20 ml of 4M HNO_3. Add 80 mg of sodium nitrite in water and stir the solution again for 15 minutes. Pass the solution at 0.25 ml·min^{-1} through a chromatographic column filled with the extraction slurry B, wash with 50 ml of 4M HNO_3, and elute plutonium with a solution 6M in hydrochloric acid and 0.02M in ammonium iodide at the same flow rate. Evaporate to dryness. Electroplate from ammonium sulphate [19].

2.4. Americium purification

The procedure for americium determination in sediment samples is similar to that described by Bojanowski et al. [11] except for a HDEHP extraction step which substitutes two anion-exchange steps. To 50 g of sediment samples 2 dis/min of ^{242}Pu, 7 dis/min of ^{241}Am, and 50 mg of Nd are added. After leaching, valence adjusting, and TOPO extraction, the plutonium analysis can be carried out from the organic phase as described in the previous section. The aqueous solution containing transplutonium elements and lanthanides is processed as follows. Evaporate almost to dryness, then dilute to 2 ltr with distilled water and adjust the pH-value to 2 with ammonia. Add oxalic acid to make the solution 5%, then control the pH-value, which should not increase above 2. Let the precipitate settle overnight. Filter, wash with a 5% oxalic acid solution and disssolve the precipitate in 8M HNO_3. Dilute to 1 ltr with water, adjust the pH-value to between 1 and 2 and add 75 g solid oxalic acid to precipitate oxalate again. In this way a good separation from iron can be achieved. Filter and wash as previously described and metathesize the oxalates to hydroxides by addition of 5 ml of 10M NaOH. Centrifuge, dissolve in the minimum volume of concentrated nitric acid, dilute to about 150 ml and adjust the pH-value to 2. Pass the solution at 0.5 ml·min^{-1} through a chromatographic column filled with the extraction slurry C, wash with 30 ml of 0.01M HNO_3 and elute americium with 70 ml of 4M HNO_3. This separation with HDEHP is necessary to eliminate traces of other transuranics,

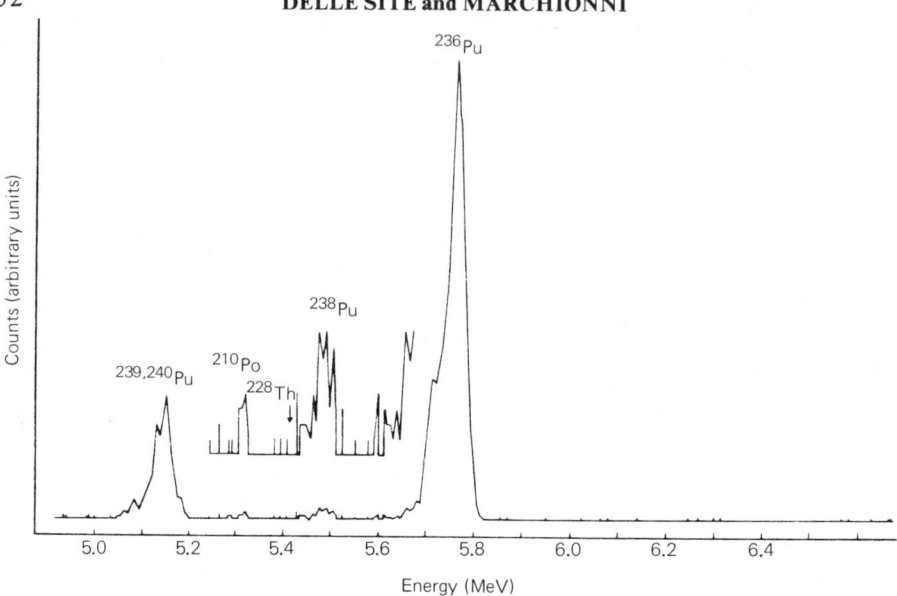

FIG.1.   α-spectrum of plutonium from 100 g dried sediment (counting time: 6000 min).

especially thorium. Evaporate to dryness and dissolve the residue in 1 to 2 ml of
1.5M HCl, add 2 ml warm 6M americium thiocyanate, neutralize with 1N NH$_4$OH
until the pink colour just disappears. Restore the colour by dropwise addition
of 0.2M HCl and pass the solution through the anion-exchange column. Wash
the column with 150 ml of 2M NH$_4$SCN containing 0.2% hydroxylamine
hydrochloride to separate the lanthanides. Elute the americium with 70 ml of
4M HCl, collecting the eluate in a beaker containing 15 ml of 16M HNO$_3$ under
stirring. Evaporate to dryness and repeat the anion-exchange procedure. Electroplate from ammonium sulphate [19].

3.   RESULTS AND DISCUSSION

The method proposed for plutonium determination has the following
characteristics. The overall recovery is quite good for sea water and marine
organisms (62.6 ± 9.7% and 81.7 ± 4.5%, respectively) and is less satisfactory for
sediments (45.4 ± 9.6%). The decontamination factors reported by Testa and
Santori [20] for single-step extraction chromatography of Pu(IV) in 4M HNO$_3$
by the Microthene-TOPO procedure are: $> 10^4$ for radium, 4×10^3 for americium
and curium, 2×10^3 for protactinium, 6×10^2 for uranium, 3×10^2 for neptunium,
1×10^2 for thorium, and 30 for polonium. Because of the significant quantity
of thorium which may be present in some environmental samples, a second
extraction step is often necessary, especially in the case of sediments. Figure 1

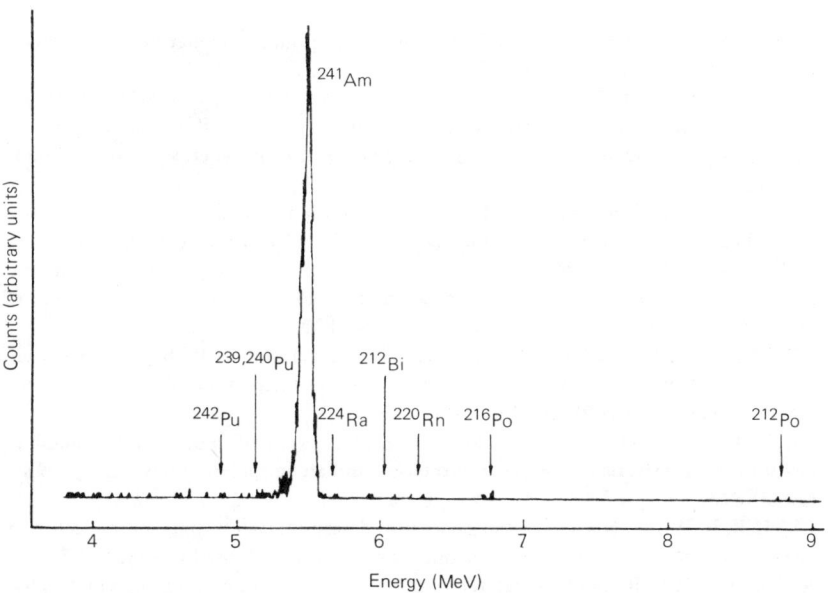

FIG.2. α-spectrum of americium from 50 g dried sediment (counting time: 4000 min;
7 dis/min ^{241}Am added).

shows the α-spectrum of a sediment sample after two extraction steps. Also, for
marine organisms two extraction steps are often necessary. The values of the
reagent blank activity are low enough to allow the detection of some femtocuries
of plutonium per sample. The method was checked by analysing some sea water
and sediment reference samples prepared by the IAEA Marine Radioactivity
Laboratory (Monaco) for intercomparison programs (SW-I-3 and SD-B-3,
respectively). The agreement between the data reported by the IAEA and the
experimental values obtained here was fairly good.

The experiments performed to test the method for americium analysis in
environmental samples gave satisfactory results. Figure 2 shows a typical spectrum
obtained with a 50 g sediment sample added with ^{242}Pu and ^{241}Am. The overall
recovery ranges between 25 and 35% and the procedure of decontamination from
plutonium and thorium seems promising.

REFERENCES

[1] LIVINGSTON, H.D., MANN, D.R., BOWEN, V.T., "Analytical procedures for transuranic
     elements in sea water and marine sediments", Analytical Methods in Oceanography, Adv.
     Chem. Ser. No.147 (1975) 124.
[2] WONG, K.M., Radiochemical determination of plutonium in sea water, sediments and
     marine organisms, Anal. Chim. Acta **56** (1971) 355.

[3] PILLAI, K.C., SMITH, R.C., FOLSOM, T.R., Plutonium in the marine environment, Nature **203** (1964) 568.

[4] BALLESTRA, S., HOLM, E., FUKAI, R., "Low-level determination of transuranic elements in marine environmental samples", Determination of Radionuclides in Environmental and Biological Materials (Symp. Central Electricity Generating Board, London, 1978).

[5] HOLM, E., FUKAI, R., Method for multi-element alpha-spectrometry of actinides and its application to environmental radioactivity studies, Talanta **24** (1977) 659.

[6] SAKANOUE, M., NAKAMURA, M., IMAI, T., "Determination of plutonium in environmental samples", Rapid Methods for Measuring Radioactivity in the Environment (Proc. Symp. Neuherberg, 1971), IAEA, Vienna (1971) 171.

[7] STATHAM, C., MURRAY, C.N., Radiochemical Separation of Plutonium, Americium and Curium from Environmental Material by Solvent Extraction. A Preliminary Report, Rapp. Comm. int. Mer Médit. **23** (1976) 163.

[8] HAMPSON, B.L., TENNANT, D., Simultaneous determination of actinide nuclides in environmental materials by solvent extraction and alpha spectrometry, Analyst **98** (1973) 873.

[9] LEVINE, H., LAMANNA, A., Radiochemical determination of plutonium-239 in low-level environmental samples by electrodeposition, Health Phys. **11** (1965) 117.

[10] AARKROG, A., "Radiochemical determination of plutonium in marine samples by ion exchange and solvent extraction", Reference Methods of Marine Radioactivity Studies II, Technical Reports Series No. 169, IAEA, Vienna (1975) 191.

[11] BOJANOWSKI, R., LIVINGSTON, H.D., SCHNEIDER, D.L., MANN, D.R., "A procedure for analysis of americium in marine environmental samples", ibid., 87.

[12] FUKAI, R., HOLM, E., BALLESTRA, S., "Modified procedure for plutonium and americium analysis of environmental samples using solvent extraction", IAEA (Monaco) Rep. No. 187 (1976) 75.

[13] EDGINGTON, D.N., ALBERTS, J.J., WAHLGREN, M.A., KARTTUNEN, J.O., REEVE, C.A., "Plutonium and americium in Lake Michigan sediments", Transuranium Nuclides in the Environment (Proc. Symp. San Francisco, 1975), IAEA, Vienna (1976) 493.

[14] HOLM, E., FUKAI, R., Determination of americium and curium by using ion-exchanges in nitric acid/methanol medium for environmental analysis, Talanta **23** (1976) 853.

[15] TESTA, C., DELLE SITE, A., The application of extraction chromatography to the determination of radionuclides in biological and environmental samples, Fifteen years' experience at the Casaccia Nuclear Centre, J. Radioanal. Chem. **34** (1976) 121.

[16] DELLE SITE, A., IANELLI, S., MARCHIONNI, V., TRIULZI, C., Preliminary results on ^{239,240}Pu and ^{238}Pu in some environmental samples of the Taranto Gulf, Rapp. Comm. int. Mer Médit. **25/26** (1979) 5.

[17] STARY, J., Separation of transplutonium elements, Talanta **13** (1966) 421.

[18] DALTON, J.C., "A new method for the determination of ^{241}Am in urine", Determination of Radionuclides in Materials of Biological Origin (Symp. AERE, Harwell, 1967).

[19] TALVITIE, N.A., Electrodeposition of actinides for alpha spectrometric determination, Anal. Chem. **44** (1972) 280.

[20] TESTA, C., SANTORI, G., Un metodo sensibile per la determinazione di bassi livelli di plutonio nelle urine di lavoratori professionalmente esposti, Giorn. Fis. San. Radioprot. **16** (1972) 1.

A METHOD FOR DETERMINING CARBONATE COMPLEXES

K. NILSSON, B. SKYTTE JENSEN
Risø National Laboratory,
Roskilde,
Denmark

Abstract

A METHOD FOR DETERMINING CARBONATE COMPLEXES.

This paper shows the value of liquid/liquid partitioning to determine the formation constants of bicarbonate and carbonate complexes of radionuclides. These constants are important to evaluate the rate of migration of complexed radionuclides that are in carbonate-saturated ground water. Experimental difficulties are discussed and reduced to provide laboratory approximations to these constants.

Radioactive waste contains long-lived radionuclides. In the event that some of the radionuclides are released it is generally accepted that the ground water is the means of transportation of the radioactivity to the biosphere, and that the migration will be the result of formation of soluble complexes between the radioactive cations and anions present in the ground water.

The composition of ground water varies, but anions such as hydroxide, chloride, sulphate, fluoride, bicarbonate and carbonate are always present. These anions form soluble complexes with the radionuclides. The physico-chemical behaviour of such soluble complexes is sufficiently well described for use in estimation of their migration behaviour, except for the soluble complexes between radionuclides and bicarbonate and carbonate. The formation constants of these complexes thus are of major importance in order to enable evaluation of the effect of carbonate complexation on migration phenomena.

Experimental difficulties are the reasons for the lack of reliable values for the stability constants. Laboratory studies of carbonate complexes have been complicated by the low solubility of most carbonates, which leaves only small concentrations of complexes in solution. The resulting dilute solutions are not easily studied and, consequently, the available data are few and often of dubious value.

It is, however, possible to adopt the method of liquid/liquid partition[1] for the determination of formation constants of bicarbonate and carbonate complexes of radionuclides.

---

[1] See ROSSOTTI, F.J.C., ROSSOTTI, H., The Determination of Stability Constants, Mcgraw-Hill Book Company, Inc. (1961).

In the method of liquid/liquid partition, two immiscible solvents containing a complexing ligand and the radionuclide to be studied are mixed until complete equilibration. Then the distribution of the radionuclide between the two solvents is measured as a function of the ligand concentration. The data thus obtained are correlated with a theoretical model and the desired formation constants obtained. The method requires only tracer amounts of the cation when radioisotopes are used due to the very sensitive detection of the metal ion. When tracer amounts are used, then each species contains only one metal ion making theoretical treatment easier and, in addition, solubility problems are minimized because experiments can be carried out in very dilute solutions.

When a radionuclide M and the ligand L are mixed, a series of complexes, ML, ML_2, ML_3, . . . may be formed in a stepwise manner. Each step is characterized by an equilibrium constant, K_1, K_2, K_3, . . . or, more commonly, the equilibrium is expressed as a function of the starting materials: M and L, and the formation products, $\beta_1, \beta_2, \beta_3, \ldots$,

$$\text{where } \beta_n = \prod_0^N K_n.$$

$$M \quad + \quad L \rightleftharpoons ML \qquad [ML] \quad = \quad K_1 [L] [M] \qquad = \beta_1 [L] [M]$$

$$ML \quad + \quad L \rightleftharpoons ML_2 \qquad [ML_2] \quad = \quad K_2 [L] [ML] \qquad = \beta_2 [L]^2 [M]$$

$$ML_2 \quad + \quad L \rightleftharpoons ML_3 \qquad [ML_3] \quad = \quad K_3 [L] [ML_2] \qquad = \beta_3 [L]^3 [M]$$

The total concentration of the radionuclide C_M in the system is the sum of the free metal ion and the various complexes, all carrying different charges of which one of the charges must be zero.

$$C_M = M^{+m} + ML^{+(m-1)} + ML_2^{+(m-2)} + \ldots \ldots$$

In the two-phase liquid system the uncharged complex will distribute itself between the organic phase and the aqueous phase whereas all the charged molecules will remain in the aqueous phase.

$$(K_d)_M = \frac{[ML_c]^{\text{organic}}}{\left(\displaystyle\sum_0^N [ML_n] \right)^{\text{aqueous}}} = \frac{\lambda \beta_c [L]^c}{\displaystyle\sum_0^N \beta_n [L]^n} \qquad (1)$$

$(K_d)_M$ is the distribution coefficient of the radionuclide; λ is the distribution coefficient of the uncharged molecule between the organic and the aqueous phase.

With carbon dioxide present in the gas phase during an experiment, the following equilibrium comes about:

$$CO_2(g) + H_2O \rightleftharpoons \underset{K_1}{H_2CO_3} \rightleftharpoons \underset{K_2}{H^+ + HCO_3^-} \rightleftharpoons \underset{K_3}{2\,H^+ + CO_3^{2-}}$$

The Henderson-Hasselbach equation can be applied, and the concentration of bicarbonate as well as carbonate can be expressed exclusively as functions of the carbon dioxide pressure and the hydrogen ion concentration

$$[HCO_3^-] = K_1 K_2\, pCO_2/[H^+]$$

$$[CO_3^{2-}] = K_1 K_2 K_3\, pCO_2/[H^+]^2$$

It is possible by carrying out the experiments at a known constant carbon dioxide pressure and by determining the pH of the experiments to calculate the exact experimental concentrations of bicarbonate and carbonate.

Titration at the experimental conditions of 1M KNO_3 and varying pCO_2 have resulted in the following values for $K_1 K_2$ and $K_1 K_2 K_3$:

$$K_1 K_2 = 10^{-7.56} \qquad K_1 K_2 K_3 = 10^{-17.24}$$

When bicarbonate and carbonate are present in the experimental solution, then competition for the metal ion will take place between the ligand and bicarbonate and carbonate. The result will be the addition of extra terms in the above equation for $(K_d)_M$.

$$(K_d)_M' = \frac{\lambda \beta_c [L]^c}{\sum\limits_{0}^{N} \beta_n [L]^n + \sum\limits_{1}^{I} \delta_i [HCO_3^-]^i + \sum\limits_{1}^{J} \epsilon_j [CO_3^{2-}]^j + \sum\limits_{1}^{K} \sum\limits_{1}^{R} \sigma_{kr} [L]^k [HCO_3^-]^r +} \qquad (2)$$

The experimental data consist of corresponding values of pH, the distribution coefficient K_d and the carbon dioxide pressure pCO_2. From the various hydrogen ion concentrations it is possible to establish the concentrations of bicarbonate and carbonate as well as the concentration of the anion of the complexing ligand.

The ligand chosen for the experiments is: 1-phenyl-3-methyl-4-benzoyl-pyrazolone-5, abb. benzoylpyrazolone or BP. Benzoylpyrazolone is a weak acid with a pK of 3.99. In a two-phase liquid system, the pK is shifted towards higher pH according to the equation: $pK' = pK + \log(1 + \gamma)$. γ is the distribution coefficient for benzoylpyrazolone between the organic and the aqueous phase.

Isoamylalcohol is used as the organic phase in the current experiments, and the concentration of the ligand anion is determined by use of the Henderson-Hasselbach equation from the pK' of 6.26 for benzoylpyrazolone in a 1:1 mixture of isoamylalcohol: water, and the experimental pH.

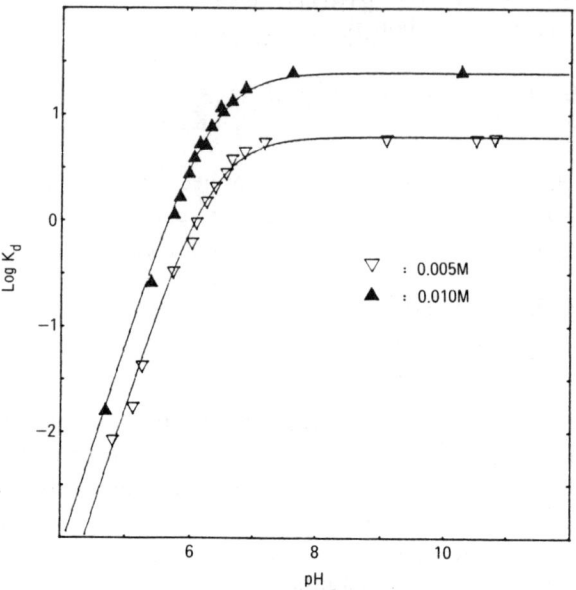

FIG.1.   The distribution of the ^{85}Sr ion in isoamylalcohol:0.1M KNO_3 at 0.005M and
0.01M of ligand concentrations with N_2 in the gas phase, as a function of pH.

The values of the stability constants are determined from Eq.(1) when
experiments are carried out in the absence of CO_2, and from Eq.(2) when CO_2
is present. In order to facilitate computations the equations are inverted and
the various coefficients, each of which represents one complexing species, are
considered as single constants. The method of least squares is then used to
establish which constants will result in the best possible fit between the
experimental values and the theoretical values. A system of simultaneous
linear equations is the result, and they are solved by the method of Gauss elimination.

Experiments at two different concentrations of benzoylpyrazolone were
carried out in an atmosphere of nitrogen using the isotope ^{85}Sr at a concentration
of 10^{-7}M. The results are presented in Fig.1. The anticipated reaction scheme is:

$$Sr^{2+} + BP^- \rightleftharpoons Sr(BP)^+ + BP^- \rightleftharpoons Sr(BP)_2$$

and using Eq.(1) the distribution coefficient is:

$$(K_d)_M = \frac{\lambda\beta_2 L^2}{(1 + \beta_1 L + \beta_2 L^2)}$$

TABLE I. CALCULATED DISTRIBUTION COEFFICIENTS FOR ^{85}Sr IN
TWO CONCENTRATIONS OF BENZOYLPYRAZOLONE

| Molar concentration of ligand | Coefficients | | | Error-sum |
	a $\times 10^{-6}$	b $\times 10^{-4}$	c $\times 10^{-2}$	
	4.39	−2.06	2.20	6.89
0.005	4.20	—	−1.46	1.73
	4.03	—	—	1.30
	5.17	−3.65	2.32	421
0.10	4.32	—	−0.79	33
	4.03	—	—	11

During computation the three constants: $a = 1/\lambda\beta_2$; $b = \beta_1/\lambda\beta_2$; $c = 1/\lambda$;
are evaluated. A minimum value for the error-sum defined as: error-sum =
Σ (Experimental data — Theoretical data)^2 is used as an indication of the best
possible fit.

The results are shown in Table I. A negative value of a constant is considered
as a mathematical artefact. Chemically it suggests that the concentration of
that particular species is insignificant. The negative term is excluded from the
theoretical model and the computations are repeated. The ultimate result of
the fitting procedure at both ligand concentrations was the finding that only
the coefficient a retains a positive value. a represents the constant $1/\lambda\beta_2$;
thus the product: $\lambda\beta_2$ is equal to 2.45×10^5.

The solid line in Fig.1 is a graphic presentation of the best possible fit
between the model and the experimental data.

The logarithmic expression of Eq.(1):

$$\log (K_d)_M = \log (\lambda\beta_c) + c \log L - \log \sum_{0}^{N} \beta_n L^n$$

can be used for a graphic presentation. At small ligand concentrations there
exists a linear relationship between $\log (K_d)_M$ and $\log L$. With increasing
concentration of the ligand, the last term takes on influence and the straight
line will curve until a condition is reached when all of the ligand is present
exclusively as the uncharged complex.

In the current experiments the ligand concentration is limited by the
solubility of benzoylpyrazolone in the organic solvent. The findings of the

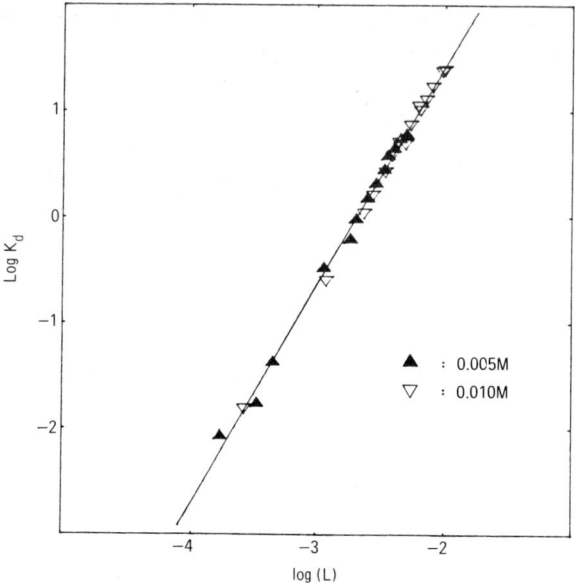

FIG.2.   *The distribution of the* [85]*Sr ion as a function of 0.005M and 0.01M of ligand concentration.*

fitting procedure presented in Table I indicate that the experimental ligand concentrations lead to results which fall on the linear portion of the curve.

In order to verify this conclusion the data of Table I were fitted to the straight-line relationship:

$$\log (K_d)_M = \log (\lambda \beta_c) + c \log L$$

using the method of least squares. Figure 2 shows that the relationship between log L and log $(K_d)_M$ is indeed linear following the equation:

$$\log (K_d)_M = 5.54 + 2.06 \log L$$

The slope of the straight line is 2.06 confirming that no complexes remain in the aqueous phase. The final product has two ligands bound to the central atom: $Sr(BP)_2$. The product $(\lambda \beta_2)$ is equal to 3.48×10^5, which agrees reasonably with the value of 2.45×10^5 found through the computations.

The experimental conditions prohibit an individual determination of the two constants, but these values are not mandatory for the determination of the bicarbonate and carbonate complexing constants.

When the equilibrium between strontium and the benzoylpyrazolone anion arises in an atmosphere of carbon dioxide then certain concentrations of bicarbonate and carbonate will exist, and competition for the strontium will take place between the ligand and bicarbonate and/or carbonate. The following reaction scheme describes the possible pathways:

$$BP^- + Sr^{2+} \rightleftharpoons Sr(BP)^+ + BP^- \rightleftharpoons Sr(BP)_2$$

$$HCO_3^- + Sr^{2+} \rightleftharpoons Sr(HCO_3)^+ + HCO_3^- \rightleftharpoons Sr(HCO_3)_2$$

$$+$$

$$BP^-$$

$$H_2CO_3 \qquad\qquad \Updownarrow$$

$$Sr(HCO_3)(BP)$$

$$CO_3^{2-} + Sr^{2+} \rightleftharpoons SrCO_3$$

At equilibrium, the reaction mixture can theoretically consist of the following strontium-containing species: Sr^{2+}, $Sr(BP)^+$, $Sr(BP)_2$, $Sr(HCO_3)^+$, $Sr(HCO_3)(BP)$ and $SrCO_3$. The uncharged organic complex, $Sr(BP)_2$ is readily extracted into the organic phase, whereas the mixed complex, $Sr(HCO_3)(BP)$ is much less likely to be found in the organic phase. All the remaining charged molecules will reside in the aqueous phase.

The uppermost pathway in the diagram describes the stepwise formation of strontium (benzoylpyrazolone)$_2$, and this is the pathway which applies when the experiments are carried out in an atmosphere of nitrogen. The results from the nitrogen experiments showed that the experimental conditions are such that strontium itself is present in the aqueous phase but no strontium-containing complexes. The final product $Sr(BP)_2$ is extracted into the organic phase.

There remain theoretically the following intermediates for formation of strontium-containing complexes,

(1) $Sr^+ + Sr(HCO_3)^+ + Sr(HCO_3)_2$

(2) $Sr^+ + Sr(HCO_3)^+ + Sr(HCO_3)(BP)$

(3) $Sr^+ + SrCO_3$

Pathway 1 assumes a stepwise formation of strontium bicarbonate; pathway 2 assumes the stepwise formation of a mixed ligand-bicarbonate complex; pathway 3 assumes the formation of strontium carbonate.

TABLE II. CALCULATED DISTRIBUTION COEFFICIENTS FOR ^{85}Sr
IN TWO PARTIAL PRESSURES OF CARBON DIOXIDE

pCO$_2$	Pathway No.	Coefficients					Error-sum
		a $\times 10^{-6}$	b $\times 10^{-11}$	c $\times 10^{-13}$	d $\times 10^{-20}$	e $\times 10^{-20}$	
	2	12.7	108	−101			311
	2	7.23	4.45				66
0.5 atm	2	7.02		4.22			7.1
	1	8.63		−4.39	5.17		21
	1	7.74			2.80		3.1
	3	7.74				2.80	3.1
	2	8.75	9.05	−4.11			11
	2	8.37		4.06			2.6
1.0 atm	2	8.55	4.53				2.1
	1	8.82		−0.56	2.85		0.7
	1	8.76			2.54		0.4
	3	8.76				2.54	0.4

Note: 1 atm = 1.01325 $\times 10^5$ Pa.

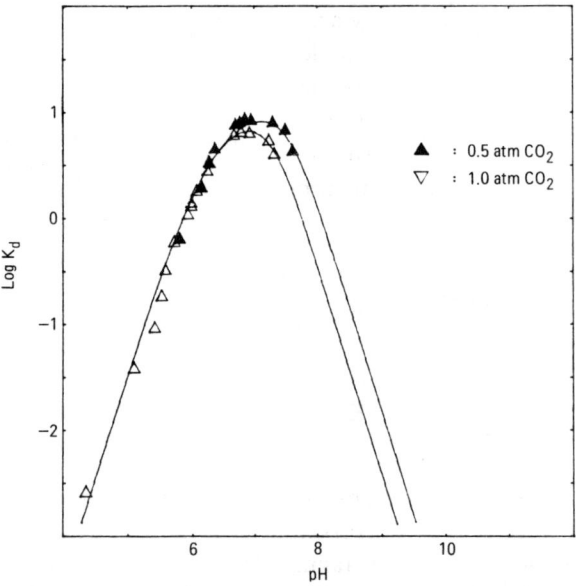

FIG.3.   The distribution of the ^{85}Sr ion in isoamylalcohol:0.1M KNO$_3$ at 0.01M of ligand
concentration with 0.5 or 1.0 atm of CO$_2$ in the gas phase, as a function of pH.

TABLE III. CALCULATED DISTRIBUTION COEFFICIENTS FOR ^{85}Sr WITH CARBONATE AND BICARBONATE COMPLEXES

| Possibility | Coefficients | | | Error-sum |
	a $\times 10^{-6}$	b $\times 10^{-20}$	c $\times 10^{-20}$	
Bicarbonate complex	8.8	2.6		6
Carbonate complex	8.3		1.6	12

Equilibration experiments were carried out using 10^{-7} M strontium and 0.01M benzoylpyrazolone at 0.5 atm and 1.0 atm of carbon dioxide. The data are presented in Fig.3, and the results of the computations are given in Table II.

A large error-sum immediately excludes pathway 2, which is the stepwise formation of a mixed ligand-bicarbonate/carbonate complex. It is also apparent from Table II by applying the previously presented reasoning that the intermediate $Sr(HCO_3)^+$ can be present only in a very low concentration. It is not possible, however, to distinguish whether $Sr(HCO_3)_2$ or $SrCO_3$ are formed. Pathways 1 and 3 are equally likely according to the error-sum.

Bicarbonate and carbonate complexes can, however, be distinguished by their dependence on the carbon dioxide pressure. There is a direct dependence if carbonate complexes are formed, whereas there is a dependence with the square of the carbon dioxide pressure if bicarbonate complexes are formed.

The results of the initial fitting procedure reduced pathways 1 and 3 to the following equations:

If $Sr(HCO_3)_2$ is formed:
$$K_d = \frac{\lambda \beta_2 L^2}{(1 + \delta_2 (HCO_3^-)^2)}$$

If $SrCO_3$ is formed:
$$K_d = \frac{\lambda \beta_2 L^2}{(1 + \epsilon_1 (CO_3^{2-}))}$$

Table III presents the results of the computation when the combined data are evaluated. The value of the error-sum clearly indicates that complex formation involving bicarbonate is taking place.

The coefficient a has a value of 8.8×10^{-6}, which means that the product $\lambda \beta_2$ is equal to 1.14×10^5. When the experiments were carried out in an atmosphere of nitrogen then $(\lambda\beta_2)$ was equal to 2.45×10^5, in very good agreement with the value of 1.14×10^5 found when carbon dioxide is also present in the gas phase.

The coefficient d is found to be equal to 2.63×10^{-20} and using the value of $(\lambda\beta_2) = 1.14 \times 10^5$, the complexity constant for the equilibrium, $Sr^{2+} + HCO_3^-$, can be calculated from the equation: $\delta_2 = (\lambda\beta_2) \times d/(K_1 K_2)^2$, as having a value of 4.

The solid lines in Fig.3 are the best possible fit between the experimental points and the model based on the theory that strontium bicarbonate is the complex formed under the experimental conditions.

SUR L'EXISTENCE
DE COMPLEXES CARBONATES DE Np(V)
EN SOLUTION AQUEUSE

A. BILLON
Service d'études analytiques,
CEA, Centre d'études nucléaires
    de Fontenay-aux-Roses,
Fontenay-aux-Roses,
France

Abstract—Résumé

ON THE EXISTENCE OF CARBONATE COMPLEXES OF Np(V) IN AQUEOUS SOLUTION.
The important amount of carbonate in the natural environment justifies an interest in the possible existence of carbonate complexes of the transuranium elements. The first results of a qualitative spectrophotometric study obtained with Np(V) are described. Judging by the shift in wavelength of the spectral bands and from their relative intensity variations, the experimental observations favour complex(es) of Np(V)-carbonate at pH 7.

SUR L'EXISTENCE DE COMPLEXES CARBONATES DE Np(V) EN SOLUTION AQUEUSE.
L'importance des carbonates dans le milieu naturel justifie que l'on s'intéresse à l'existence éventuelle de complexes carbonatés des éléments transuraniens. Les premiers résultats d'une étude spectrophotométrique qualitative obtenue avec Np(V) sont décrits. D'après la déplacement des bandes spectrales en longueur d'onde et les variations d'intensité relative de celles-ci, les observations expérimentales sont en faveur de complexe(s) Np(V)-carbonate à pH 7.

INTRODUCTION

La migration des espèces chimiques dans le sol est fonction de très nombreux facteurs, géologique, hydrodynamique, chimique, etc. Une mention particulière doit être faite aux réactions de complexation dont le produit final peut être une espèce neutre ou chargée dont la solubilité est différente de l'espèce de départ.

Les anions complexants peuvent être amenés par l'homme ou, au contraire, exister de façon naturelle comme les carbonates, chlorures, sulfates, constituants normaux à teneur variable des eaux minéralisées.

Dans le milieu naturel, l'importance des carbonates est énorme et c'est pourquoi la possibilité de l'existence de complexes carbonatés des éléments transuraniens dans les sols doit être envisagée.

Nos connaissances de ces complexes sont limitées, d'une part en raison de la chimie propre des éléments transuraniens qui rend les déterminations expérimentales longues et difficiles, d'autre part parce que le domaine de concentration très faible rencontré est un domaine où l'analyste dispose d'un arsenal de méthodes certes très sensibles, mais pas nécessairement spécifiques des états de valence.

Le présent travail est une contribution qualitative à l'étude de complexes carbonatés de Np(V) par spectrophotométrie d'absorption. Le choix de l'élément Np au degré d'oxydation (V) a été guidé par les considérations de simplicité suivantes qui s'attachent à l'ion NpO_2^+ :

a) hydrolyse faible

b) dismutation nulle dans la zone de pH étudiée

c) coefficient d'absorption molaire élevé de la bande située à 980 nm.

L'application de la spectrophotométrie à l'étude des complexes en solution est bien connue [1]. Comme, de plus, chaque degré d'oxydation du neptunium présente un spectre d'absorption caractéristique, nous avons en principe la possibilité de déceler des modifications de degrés d'oxydation de l'espèce étudiée, si elles ont lieu.

RESULTATS EXPERIMENTAUX

Comportement de Np(V) en présence d'un excès de carbonate

a) milieu acide

Des solutions à pH différents ont été préparées, pour lesquelles le rapport (carbonate)$_{total}$/(Np) = 25. Le terme (carbonate)$_{total}$ s'adresse à la somme du carbonate libre sous toutes ses formes (CO_3^{2-}, HCO_3^-, CO_2) et complexé s'il y a lieu.

La concentration de neptunium est (Np) = 8,08 × 10^{-4} M.

Le pH des solutions est ajusté par ajout de $HClO_4$ ou de NaOH. Le spectre des solutions est tracé immédiatement après préparation, sauf mention particulière.

L'examen des spectres montre qu'en milieu acide et jusqu'à pH = 5,5, c'est l'espèce NpO_2^+ qui existe seule en solution. En effet, nous n'observons dans tout le spectre que la bande d'absorption à 980 nm caractéristique de NpO_2^+. Nous avons, de plus, observé que l'absorbance de cette bande reste constante à la précision des mesures près entre pH = 0 et pH = 5,5 ($\epsilon = 400$ cm^{-1}·M^{-1}).

A pH = 5,9, nous notons une diminution de l'absorbance d'environ 8%, diminution qui s'accentue avec le temps.

A pH = 6,9, une réaction de précipitation se produit. La valeur de l'absorbance à 980 nm, mesurée sur la solution surnageante, ne vaut plus que le 1/10^{è} de ce qu'elle était en milieu acide, et un épaulement situé à 990 nm apparaît dans la bande spectrale (fig. 1). Un calcul simple montre qu'à ce pH, le produit de solubilité de l'hydroxyde NpO_2OH n'est pas atteint. L'espèce qui précipite n'est donc pas l'hydroxyde, mais un carbonate acide ou mixte; en effet, on peut supposer que la diminution de (NpO_2^+) en relation avec la diminution de l'absorbance

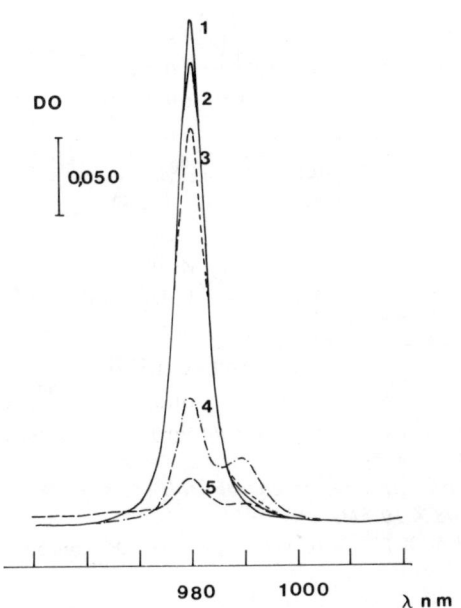

FIG.1. *Modification des spectres d'absorption de Np(V) en présence de carbonate, en fonction du pH (960 à 1020 nm).*
Courbe 1:pH=1,73 à 5,30; courbe 2:pH=5,9; courbe 3:pH=5,9 après 24 h; courbe 4:pH=7,10; courbe 5:pH=6,9.
Conditions: $(Np)_T = 8,08 \times 10^{-4} M$
$(CO_3)_T = 2 \times 10^{-2}$ (sauf courbe 4:1,60 × 10^{-2} M).

à 980 nm est due à la formation d'une espèce qui absorberait à 990 nm. Dans les conditions expérimentales, le produit de solubilité de cette espèce est atteint et c'est pourquoi nous observons:
— une réaction de précipitation
— l'absence d'un pic important à 990 nm.
A ce stade, il est difficile de préciser la nature du précipité.

b) milieu basique

Des solutions ont été préparées par ajout de la solution de neptunium à une solution équimolaire en HCO_3^- et CO_3^{2-}. $(Np)_T = 8,08 \times 10^{-4}$ M, $(CO_3)_T = 8 \times 10^{-2}$ M. Ces solutions sont à la limite de la précipitation. Celle-ci se produit dans un délai de quelques heures à quelques jours.

Nous avons représenté sur la figure 2 (courbe 2), le spectre correspondant à de telles solutions. Les bandes spectrales à 980 nm et 990 nm observées précédemment ont disparu, faisant place à deux bandes assez larges dont les maximums se situent à 997,5 nm et 972 nm.

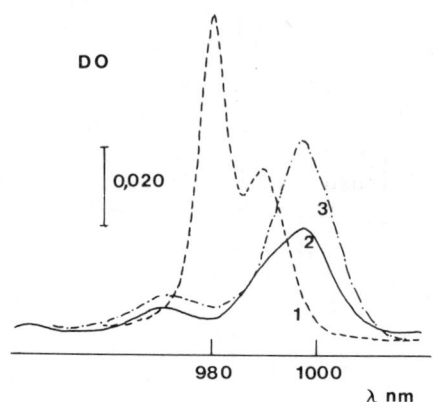

FIG.2. *Modification des spectres d'absorption de Np(V) en présence de carbonate, en fonction du pH (960 à 1020 nm).*
Courbe 1:pH=7,15; courbe 2:pH=9,75; courbe 3:pH=9,84.
Conditions: $(Np)_T = 8,08 \times 10^{-4} M.$
$(CO_3)_T = 1,60 \times 10^{-2} M$ *(courbe 1)*, $8 \times 10^{-2} M$ *(courbe 2)*, *0,16 M (courbe 3).*

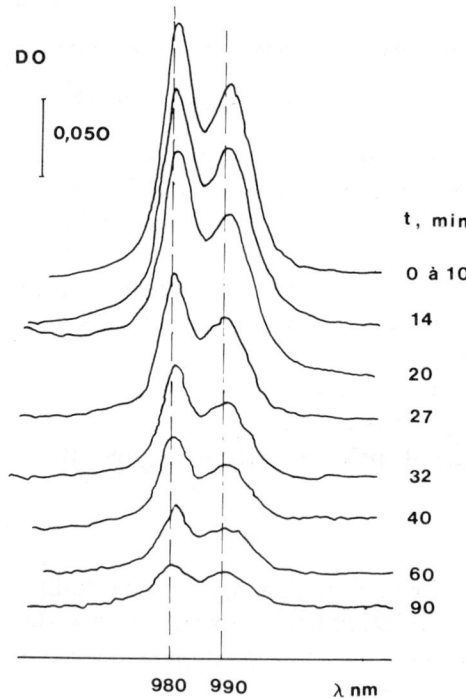

FIG.3. *Evolution des spectres d'absorption de Np(V), en présence de carbonate, en fonction du temps.*
Conditions $(Np)_T = 8,08 \times 10^{-4} M$
$(CO_3)_T = 1,6 \times 10^{-2} M$
$pH = 7,10.$

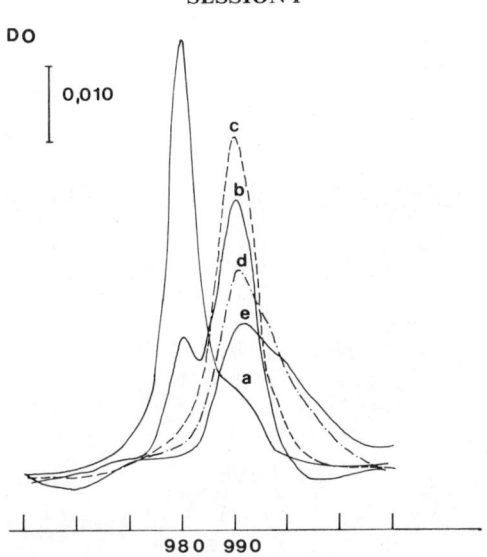

FIG.4. Modification des spectres d'absorption de Np(V), en présence de carbonate, en fonction du pH (960 à 1020 nm).
Courbe a : pH = 7,20; courbe b : pH = 7,90; courbe c : pH = 9,00;
courbe d : pH = 11,10; courbe e : pH = 11,40.
Conditions: $(Np)_T = 1,6 (2) \times 10^{-4} M$
$(CO_3)_T = 3,2 \times 10^{-3} M.$

Si l'on augmente la concentration de carbonate, par exemple $(CO_3)_T = 1,6 \times 10^{-1}$ M, la solution devient stable et nous enregistrons une absorbance plus importante aux deux longueurs d'ondes citées (courbe 3 de la figure 2).

c) facteur temps

A pH = 7, les essais effectués montrent que les solutions précédemment obtenues ne sont pas très stables. Nous avons suivi l'évolution des spectres en fonction du temps. L'intensité des *deux* bandes d'absorption à 980 et 990 nm diminue régulièrement pour ne plus valoir que quelques millièmes au bout d'une heure (fig. 3). Des essais de plus longue durée sur des solutions de pH légèrement différent n'ont pas été effectués.

d) influence de la concentration de neptunium

Nous avons cherché à utiliser des concentrations de neptunium les plus faibles possible compatibles avec la détection spectrophotométrique. Une concentration $1,62 \times 10^{-4}$ de neptunium, correspondant à une dilution 5 des solutions étudiées

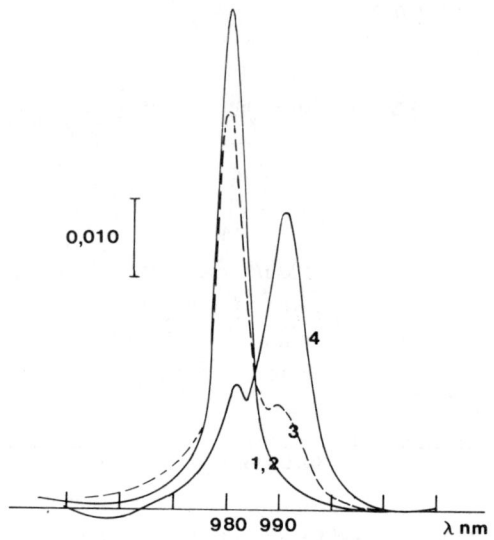

FIG. 5. *Evolution des spectres d'absorption de Np(V), en présence de carbonate, en fonction du rapport $(CO_3)_T/(Np)_T = r$.*
Courbes 1 et 2: $(Np)_T = 3,2 \times 10^{-4} M$ $r = 1$ et 2
Courbe 3: $(Np)_T = 3,2 \times 10^{-4} M$ $r = 10$
Courbe 4: $(Np)_T = 1,6 \times 10^{-4} M$ $r = 20$
pH = 7,10 ± 0,05.

en a) et b), donne en milieu non complexant une absorbance de 0,064 parfaitement mesurable lorsqu'on utilise l'échelle dilatée (0,1 d'absorbance pleine échelle) du spectrophotomètre. Nous avons choisi $(CO_3)_T = 3,2 \times 10^{-3}$ M. La figure 4 présente un ensemble de spectres observés entre pH = 7,29 et pH = 11,40. Aucune réaction de précipitation n'a été observée.

e)   *influence du rapport* $(CO_3)_T/(Np)_T$

    Le pH a été fixé à 7,10 ± 0,10; $(Np)_T = 3,23 \times 10^{-4}$ M.
    Il apparaît que l'augmentation du rapport $(CO_3)_T/(Np)_T$ se traduit par une diminution de l'intensité de la bande spectrale à 980 nm au profit de celle à 990 nm (fig. 5).

DISCUSSION

Existence d'un complexe carbonaté

Elle apparaît probable au vu des résultats précédents. La constante K_H de la réaction

$$NpO_2^+ + H_2O \rightleftharpoons NpO_2OH + H^+ \quad \text{vaut} \quad 10^{-9,5 \pm 0,5}$$

ce qui signifie qu'à pH = 9,5 ± 0,5, le neptunium en solution est présent sous forme NpO_2^+ et NpO_2OH en quantité égale (le produit de solubilité n'étant pas atteint). NpO_2OH existe donc déjà à pH = 7 en très faible quantité. On peut hésiter sur la nature de l'espèce qui absorbe à 990 nm, soit NpO_2OH, soit un complexe carbonaté de NpO_2^+. Nous avons constaté qu'une solution de $Np(V) 1,232 \times 10^{-4}$ M à pH = 8 en l'absence de carbonate présente le spectre caractéristique de NpO_2^+ ($\epsilon = 400$ cm$^{-1} \cdot$ M^{-1}). Cette solution est stable et l'hydroxyde ne précipite pas. Il semble donc que l'on doive attribuer la bande spectrale à 990 nm observée dès pH = 7 à une autre espèce que NpO_2OH, soit $NpO_2CO_3^-$ ou NpO_2HCO_3 par exemple.

En milieu basique, la bande spectrale à 997 nm est large — spectres 2 et 3 de la figure 2 — ce que l'on observe généralement dans le cas des bandes d'absorption de composé d'hydrolyse.

Il est intéressant de noter que dans les spectres d'absorption de la figure 3, l'intensité des deux bandes diminue simultanément. En effet, la diminution de l'absorbance à 980 nm est normale puisque (NpO_2^+) diminue. Nous expliquons la diminution de l'absorbance à 990 nm par la participation de l'espèce absorbant à cette longueur d'onde à un composé peu soluble résultant de l'association entre d'une part un anion contenant NpO_2^+ et, d'autre part, Na^+ présent dans la solution (formation d'un sel double).

Nous avons vérifié l'absence d'un autre degré d'oxydation du neptunium (en particulier de Np(IV) qui, lorsqu'il est présent, absorbe vers 720 nm).

Enfin, à pH constant, proche de la neutralité, une augmentation de $(CO_3)_T$ a une influence sur l'intensité relative des deux bandes d'absorption à 980 nm et 990 nm. Ce résultat — diminution de l'intensité de la bande à 980 nm avec pour corollaire augmentation de celle à 990 nm — va dans le sens de l'existence d'un complexe Np(V)-carbonate.

A notre connaissance, seule une publication [2] fait allusion à un complexe NpO_2HCO_3 en solution, en citant des valeurs de sources non publiées.

Interaction neptunium-biosphère

Les degrés d'oxydation les plus probables de Np dans la biosphère sont + IV et + V, compte tenu des potentiels redox des différents couples. Le degré V prédomine et on trouverait de préférence Np(IV) dans un milieu privé d'oxygène en présence d'éléments à caractère réducteur assez marqué.

Si ce neptunium est à l'état de traces, ce qui sera le cas, en raison de la dispersion par le milieu naturel, il se trouvera toujours en présence d'un grand excès de carbonate. La forme chimique sous laquelle il faut s'attendre à trouver Np(V) sera donc celle d'un complexe carbonaté, en l'absence évidemment d'agents complexants naturels organiques plus forts.

CONCLUSION

Cette étude des espèces de Np(V) en présence de carbonate par spectrophoto-métrie d'absorption a permis de mettre en évidence plusieurs espèces dont il conviendra de préciser la nature exacte.

La confirmation du point isobestique vers 985 nm entre NpO_2^+ et la forme complexe ainsi que la mesure du pH pour lequel celle-ci présente un maximum d'absorption devraient permettre de déterminer sinon très exactement, tout au moins avec une précision acceptable, une constante de formation.

Des composés solides tel que NpO_2HCO_3, $NpO_2M_5(CO_3)_3$, M = Na, K ou Cs, sont signalés dans la littérature [3, 4]. Ces composés ont été préparés à des fins d'étude de propriétés physiques. Ils pourraient cependant servir de point de départ pour une étude des complexes carbonatés en phase aqueuse.

REFERENCES

[1] ROSSOTTI, F.J.C., ROSSOTTI, H., The Determination of Stability Constants, Mc Graw Hill (1961).
[2] MOSKVIN, A.I., GELETSEANU, I., LAPITSKII, A.V., Dokl. Akad. Nauk. SSSR **149** 3 (1963) 611.
[3] KEENAN, T.K., KRUSE, F.H., Inorg. Chem. **3** 9 (1964) 1231.
[4] GORBENKO-GERMANOV, D.S., ZENKOVA, R.A., Russ. J. Inorg. Chem. **11** 3 (1966) 282.

DISTRIBUTION OF TRANSURANIC ISOTOPES
IN THE SURFACE WATER OF
THE NORTH SEA AND ADJACENT REGIONS

H.-F. EICKE
Deutsches Hydrographisches Institut,
Hamburg,
Federal Republic of Germany

Abstract

DISTRIBUTION OF TRANSURANIC ISOTOPES IN THE SURFACE WATER OF THE
NORTH SEA AND ADJACENT REGIONS.
　　Distribution patterns of transuranics in the North Sea area are presented. The presence
of ^{238}Pu, $^{239+240}$Pu, ^{241}Am and ^{244}Cm is described as originating mostly from three nuclear
reprocessing plants. Plutonium ratios confirm that only a small amount of these isotopes is
from fallout from nuclear weapons tests. It is suggested that the transport mechanism of
plutonium is analogous to that of ^{137}Cs.

1. INTRODUCTION

From our investigations during recent years [1–3] concerning the distribution
and spreading of ^{137}Cs in the region of the North Sea (and from those of other
authors [4–6]) we know (a) that the water which flows out of the Irish Sea through
the Minch towards the North reaches the North Sea in the vicinity of the Orkney
Islands and flows along the English East coast towards the South; and (b) that
the water which enters the North Sea through the English Channel streams along
the Belgian, Dutch, German, and Danish coasts towards the North.

As Murray, Kautsky et al. [7, 8] have shown, transuranic isotopes – in
detectable quantities – are transported into the North Sea by the ocean currents
not only from around the northernmost point of Scotland but also from the
English Channel in the South. They originate principally from nuclear fuel
reprocessing plants at Windscale near the Irish Sea, and to a far less extent from
Dounreay in the North of Scotland, as well as from La Hague in northern France.

In order to gain better knowledge concerning the behaviour of the Transuranics in the North Sea, measurements concerning the presence and the spreading
of different transuranic isotopes in the North Sea area were carried out.

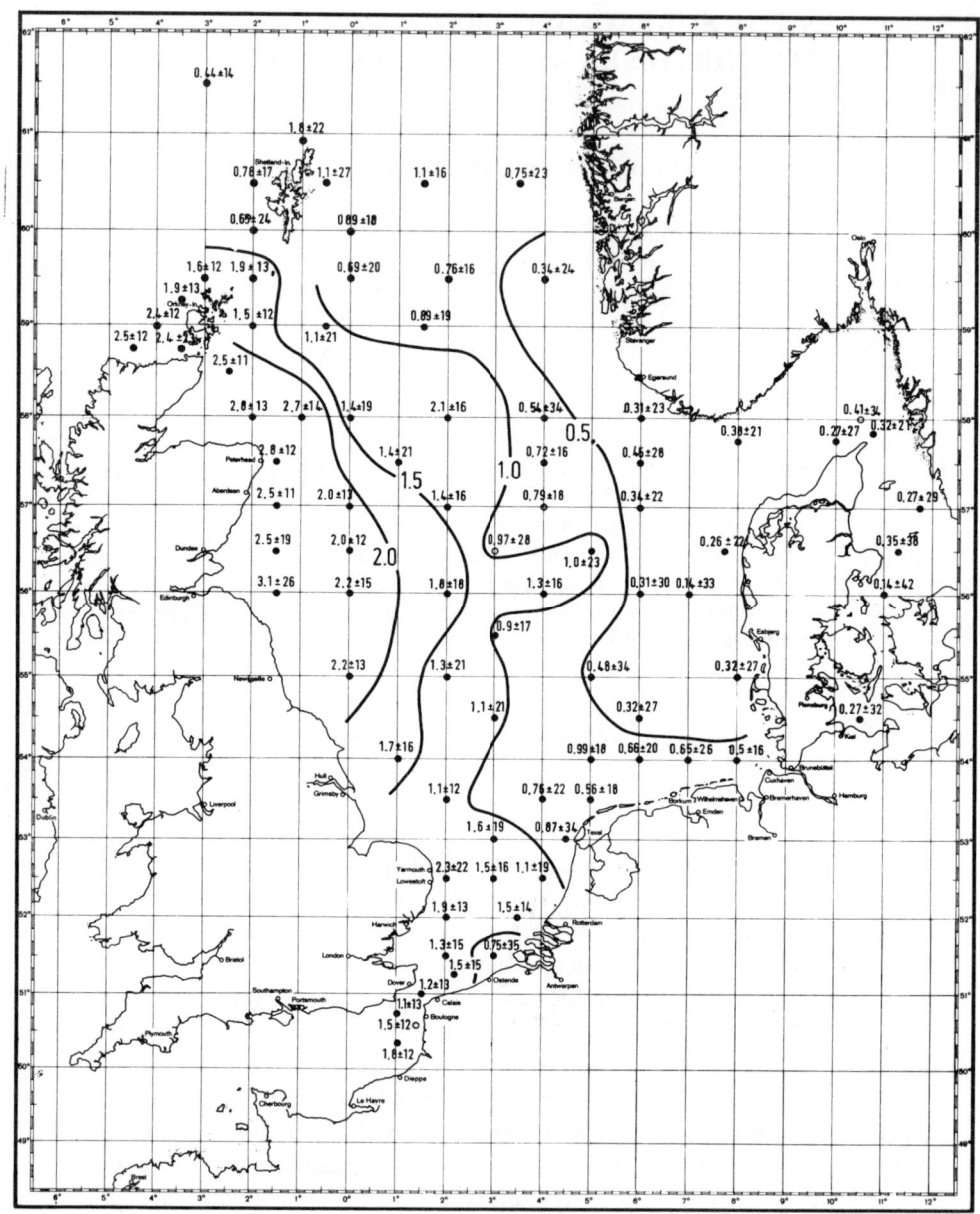

FIG.1.   Plutonium-239/240, February/March 1978.   Results in fCi/ltr.   Error terms: 1 σ
propagated error in %.

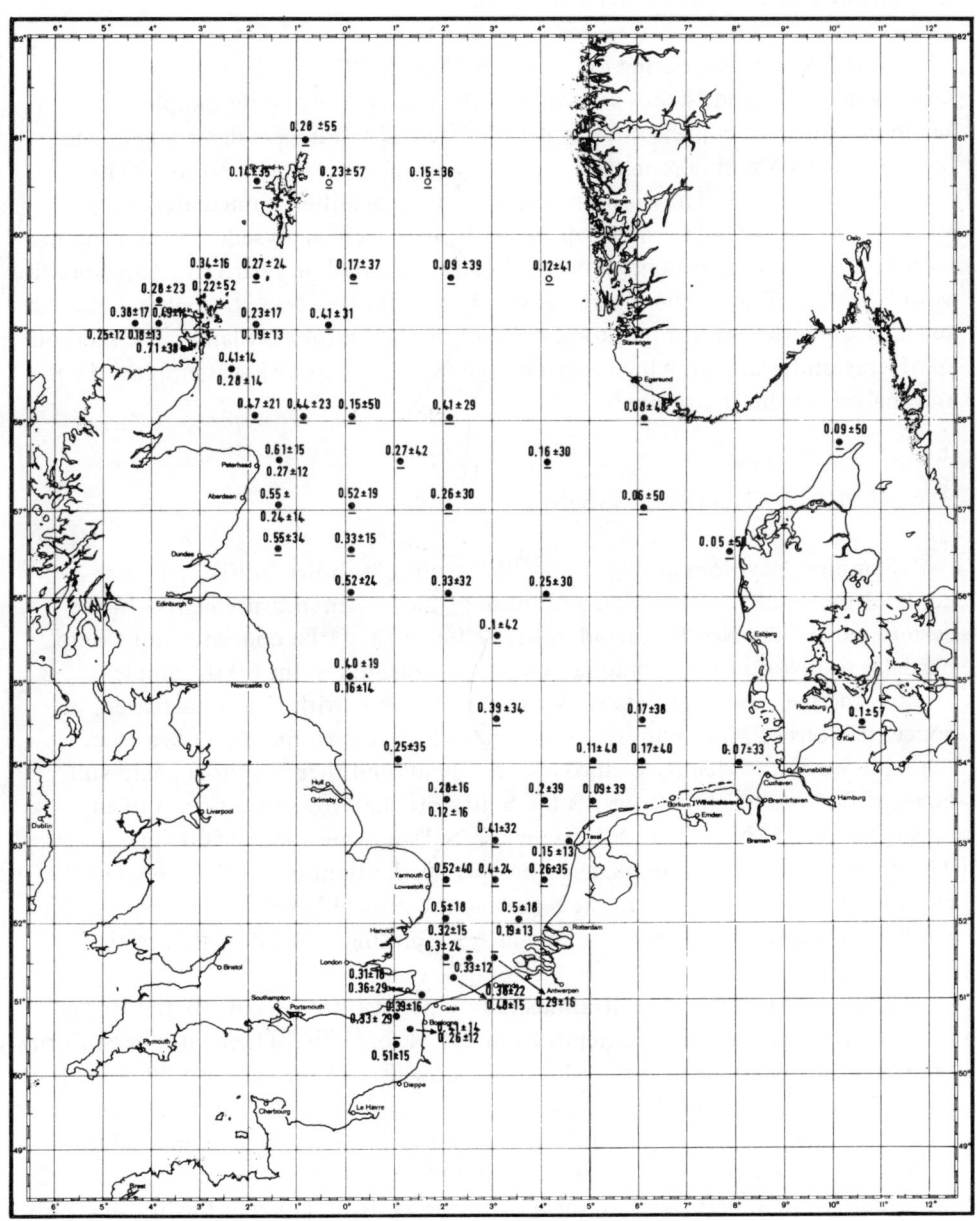

FIG.2. Plutonium-238, americium-241, February/March 1978. Results in fCi/ltr. Error terms: 1σ propagated error in %.

2.   SAMPLING AND ANALYSIS

In the North Sea, during February and March 1978, in the region of the Orkney and Shetland Islands as well as in the Kattegat, 92 water samples were taken in which the concentrations of $^{239+240}$Pu and ^{238}Pu were determined, and also that of ^{241}Am in 20 samples — whereby sample volumes of 60 to 200 ltr of sea water were used. The unfiltered samples were acidified immediately they were taken, and stored in plastic drums until they were analysed. The transuranic isotopes in part of the samples were already separated on board by coprecipitation with $Fe(OH)_3$. The further chemical separation and the α-spectrometrical measurement of the different isotopes took place in the laboratory on land. The method of Murray and Statham, which is given in detail elsewhere [9], was employed for the analysis of the transuranics.

3.   RESULTS AND DISCUSSION

Sampling positions and the $^{239+240}$Pu results are shown in Fig.1; Figure 2 contains the results of the ^{238}Pu and those of the ^{241}Am measurements. In the eastern part of the North Sea and in the Kattegat the ^{238}Pu concentrations were partially so low that they could no longer be measured in the 60 ltr samples.

As the measurement data show, water masses — with clearly increased concentrations of plutonium isotopes — have flowed into the North Sea from the Irish Sea in the vicinity of the Orkney Islands and there isotopes could still be identified on their way towards the South off the English East coast in an area between 52° N and 53° N (distance from Windscale about 1500 km). The plutonium — which originated from the northerly influx — has spread in the easterly direction especially in the region of the central North Sea to about 5° E, whereby a tongue can be recognized which is pointing in the direction of the Skagerrak.

The highest ^{239}Pu concentrations, 2.0 to 3.1 fCi/ltr*, occur off the English East coast in a region which extends from 59° N to 55° N. Therewith, in conformity with that, higher concentration values also show here in the case of ^{238}Pu (up to 0.61 fCi/ltr). If one calculates the activity ratio of ^{238}Pu/$^{239+240}$Pu from samples taken from the region quoted, this results here — on an average — in a value of 0.19±0.14, which considerably exceeds that of the fallout ratio value (0.06±0.02 according to the values given by various authors) [10, 11]. From this it can clearly be seen that the greatest part of the plutonium originates from nuclear energy plants, especially from the nuclear fuel reprocessing plants, and only to a small extent from fallout.

---

* $1 Ci = 3.70 \times 10^{10}$ Bq.

TABLE I. CURIUM CONCENTRATIONS IN SEA WATER IN THE REGION OF THE SOUTHERN NORTH SEA COAST

For comparison purposes, the concentrations of plutonium and americium are also given

Position	^{244}Cm (fCi/ltr) ± %	$^{239+240}$Pu (fCi/ltr) ± %	^{238}Pu (fCi/ltr) ± %	^{241}Am (fCi/ltr) ± %
50°35′ N, 01°25′ E	0.25 ± 12	1.49 ± 12	0.41 ± 14	0.26 ± 12
51°15′ N, 02°10′ E	0.16 ± 22	1.45 ± 15	0.38 ± 22	0.48 ± 15
51°30′ N, 03°00′ E	0.44 ± 14	0.75 ± 35	ND	0.29 ± 16
52°00′ N, 03°30′ E	0.34 ± 12	1.50 ± 14	0.5 ± 18	0.19 ± 13
53°00′ N, 04°30′ E	0.50 ± 11	0.87 ± 34	ND	0.15 ± 13
54°00′ N, 08°00′ E	0.54 ± 11	0.5 ± 16	0.7 ± 33	ND

ND: not detected.

Errors: 1 σ propagated in %.

As, in the same region, the highest ^{137}Cs concentrations (12 to 15 pCi/ltr) which have so far occurred in the North Sea were also measured, it can be conluded that this part of plutonium which reaches the North Sea from the Irish Sea — in regard to the transport mechanism — behaves analogously to the caesium.

^{241}Am has also entered the North Sea from the Pentland Firth. It was ascertained not only westwards and northwards of the Orkney Islands but also in a strip off the English East coast, and could be detected towards the South to about 53°30′ N. The concentrations lay between 0.12 and 0.28 fCi/ltr. The data material which is available to date is not sufficient to make further statements.

The waters coming from the English Channel which enter the southern North Sea contain plutonium concentrations which are lower than those measured in the North off the English East coast; that means that the main quantity of the plutonium now reaches the North Sea from the North and not from the South.

The influence of the waste waters of the nuclear fuel reprocessing plant in La Hague can be recognized by the activity ratio of ^{238}Pu/$^{239+240}$Pu of about 0.27±0.06 in the Channel region. The further penetration of plutonium in the southern coastal region of the North Sea can be traced as far as into the German Bight. Here also one can deduce, from the ratio of these plutonium isotopes (>0.14), that the incidence of the plutonium originates from fallout only to a small extent. Off the Danish coast and in the Kattegat there occurred only very low $^{239+240}$Pu concentrations (<0.41 fCi/ltr) and, in fact, ^{238}Pu concentrations were — in part — so low that they could not be measured. For that reason, to date it is not possible to make any further statements about it.

Americium-241 could be identified from the Straits of Dover along the coast as far as the island of Texel; however, it did not occur any more in the German Bight. The transport route of about 600 km from the nuclear fuel reprocessing plant at La Hague to Texel appears to be relatively short compared with the distance of about 1500 km which the americium — originating from Windscale — had covered as far as the position 53°30' N, 2° E. It is possible that the americium coming from the Channel was more quickly removed from the water — in the shallow water areas with high suspended matter content off the coast, above all off the mouth of the Rhein — whereby it was adsorbed by particulate substances and sedimented.

The ^{244}Cm concentrations found in the coastal region of the southern North Sea are given in Table I.

The appearance of the highest ^{244}Cm concentrations off Texel (0.50 fCi/ltr) and in the German Bight (0.54 fCi/ltr), at first sight, appears to be unusual. However, if one takes into account that the water body found at the end of April 1977 off Ostende and off Europoort with 0.75 and 0.78 fCi/ltr respectively required about 300 days to reach the German Bight (equivalent to a distance of about 300 nautical miles and a transport velocity of about 1 nautical mile per day, according to Kautsky [1]) than the ^{244}Cm concentrations found in the German Bight at the beginning of March 1978 could originate from this water body. If this assumption is correct, then it means that curium can also be carried over large distances in sea water and that, with reference to its transport behaviour, it is possibly more similar to that of plutonium than to that of americium. As, however, the form in which the curium in sea water is present is unknown and certain knowledge concerning the transport behaviour is not available, further investigations must be undertaken in order to be able to decide if the speculative view, referred to above, is correct.

REFERENCES

[1] KAUTSKY, H., Dtsch. Hydrogr. Z. 26 (1973) 241.
[2] KAUTSKY, H., Dtsch. Hydrogr. Z. 29 (1976) 217.
[3] MURRAY, C.N., KAUTSKY, H., Est. Coast. Mar. Sci. 5 (1977) 319.
[4] JEFFERIES, D.F., PRESTON, A., STEELE, A.K., Mar. Pollut. Bull. 4 (1973) 118.
[5] MITCHELL, N., Fisheries Radiobiol. Laboratory, Techn. Rep. 12 (1977) 32.
[6] LEE, A., Oceanography and Marine Biology (1970) 8.
[7] MURRAY, C.N., KAUTSKY, H., HOPPENHEIT, M., DOMIAN, M., Nature 276 (1978) 225.
[8] MURRAY, C.N., KAUTSKY, H., EICKE, H.-F., Nature 278 (1979) 617.
[9] MURRAY, C.N., STATHAM, G., Dtsch. Hydrogr. Z. 29 (1976) 69.
[10] HARDY, E.P., KREY, P.W., VOLCHOK, H.L., Nature 241 (1973) 444.
[11] MURRAY, C.N., EICKE, H.-F., Dtsch. Hydrogr. Z. 30 (1977) 1.

VERTICAL DISTRIBUTION OF TRANSURANIC NUCLIDES IN THE EASTERN MEDITERRANEAN SEA

R. FUKAI, S. BALLESTRA, M. THEIN
International Laboratory of Marine Radioactivity, IAEA,
Musée Océanographique,
Principality of Monaco

Abstract

VERTICAL DISTRIBUTION OF TRANSURANIC NUCLIDES IN THE EASTERN
MEDITERRANEAN SEA.

The results of measurements on ^{238}Pu, $^{239+240}$Pu and ^{241}Am in sea water collected from
vertical profiles at stations located in the eastern Mediterranean are presented and compared with
those in the western Mediterranean basin. The results confirm that the different vertical distribution patterns between plutonium and americium observed in the western basin are extended to
the eastern basins. The ^{241}Am/$^{239+240}$Pu activity ratios tend also to be lower in near surface
layers (~ 0.1) and higher in deep layers ($\geqslant 0.3$). In the deep layers at a station off Egypt
extremely high ratios exceeding 0.8 were observed. These high ratios are considered to indicate
effects of suspended matter carried down by the Nile River to the Mediterranean. Comparisons
of estimated fallout delivery of ^{137}Cs and $^{239+240}$Pu to the Mediterranean with their water column
inventories suggest that the major fractions of these isotopes delivered are still in the water column.

1. INTRODUCTION

While the vertical distribution of transuranic nuclides such as ^{238}Pu, $^{239+240}$Pu
and ^{241}Am has been studied in the Atlantic [1, 2] and the Pacific [2, 3] as well as
in the western Mediterranean [4, 5], no data have been available until now from
the eastern Mediterranean basins. Since it is known that the special hydrographical
and meteorological conditions prevailing in the northwestern Mediterranean cause
intermittent vertical water mixing [6], the systematic difference in the vertical
distribution observed between plutonium and americium in this area [5] is not
necessarily regarded as a specific feature characterizing the whole Mediterranean
Sea. The measurements of transuranic nuclides, similar to those made in the
northwestern Mediterranean, have been conducted on sea water samples collected
from vertical profiles (down to 2000 m) located in the eastern Mediterranean Sea,
the Tyrrhenian, Ionian and Levantine basins. The results of these measurements
are presented below and compared with those obtained from the northwestern
Mediterranean.

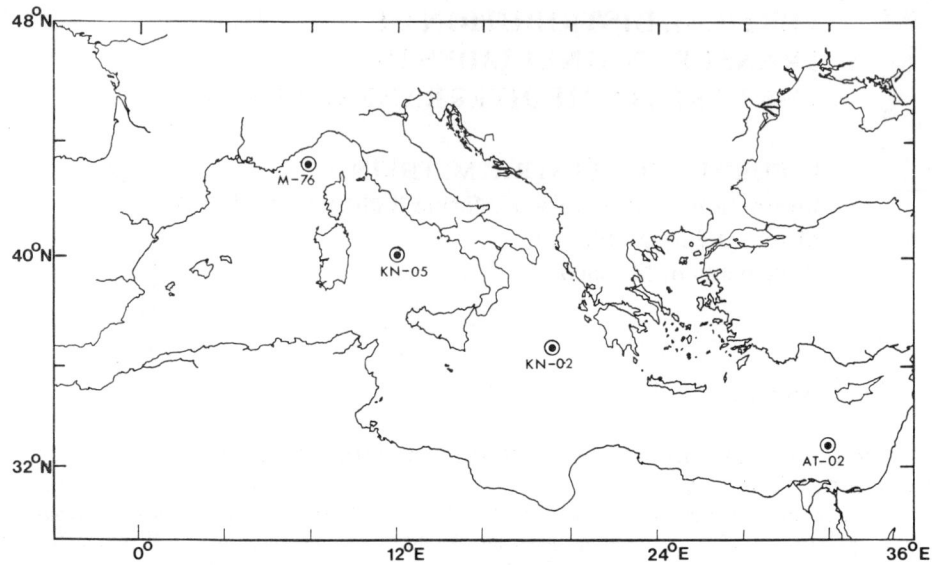

FIG.1.   Sampling stations in the Mediterranean Sea.   St. M-76 represents a station in the northwestern Mediterranean (off Monaco), where previous studies were carried out.

2.   MATERIAL AND METHODS

The sea water samples were collected during the cruises conducted by USNS "Kane" and R.V. "Atlantis II" in April 1977 across the Mediterranean from the stations illustrated in Fig. 1. This figure also includes a profile station in the northwestern Mediterranean (St. M-76), where similar studies were previously carried out [5].

Approximately 200 ltr of sea water were collected with several Niskin bottles (35 ltr) from each depth, acidified immediately to pH ~ 1.5 without filtration and transported to the laboratory at Monaco.   The radiochemical procedures described by Ballestra et al. [7] were adopted for the measurements of ^{238}Pu, $^{239+240}$Pu and ^{241}Am. In these procedures the separation of americium was achieved by the anion-exchange sorption from the nitric acid/methanol medium and successive differential elution of americium [8]. Alpha spectrometry of separated transuranic nuclides was carried out by using silicon surface barrier detectors coupled with ORTEC pulse-height analysers. Since the intervals between the sample collection and the analysis exceeded more than a year in all cases, the results of ^{241}Am measurement were corrected for the ingrown ^{241}Am from the decay of ^{241}Pu during storage, assuming that the ^{241}Pu/$^{239+240}$Pu activity ratio in the samples had been 7.0 when they were collected, the value which is the average activity ratio for integrated global fallout [9].

TABLE I. VERTICAL DISTRIBUTIONS OF ^{137}Cs, ^{238}Pu, $^{239+240}$Pu AND ^{241}Am IN THE TYRRHENIAN, IONIAN AND LEVANTINE BASINS

Station	Depth (m)	Cl (°/$_{oo}$)	^{137}Cs* (fCi/ltr)	^{238}Pu* (fCi/ltr)	$^{239+240}$Pu* (fCi/ltr)	^{241}Am* (fCi/ltr)
No. KN-05	S**	21.03	119±6	0.18 ±0.02	1.12±0.07	-
Coord.: 39°59'N	100	21.13	120±6	0.12 ±0.02	1.6 ±0.1	-
17°01'E	250	21.28	137±5	0.06 ±0.01	1.51±0.08	-
(Tyrrhenian Sea)	500	21.41	92±3	0.05 ±0.01	1.58±0.08	-
Date of Coll.:	1000	21.35	26±1	0.03 ±0.01	0.66±0.04	-
15 April 1977	1500	21.31	7±1	0.01 ±0.01	0.38±0.03	-
Depth: 3577 m	2000	21.29	11±1	0.02 ±0.01	0.18±0.01	-
No. KN-02	S**	21.33	114±7	0.07 ±0.01	1.28±0.08	0.087±0.009
Coord.: 37°00'N	100	21.41	182±4	0.059±0.009	1.44±0.07	0.07 ±0.01
18°59'E	250	21.49	126±4	0.057±0.008	1.6 ±0.1	0.16 ±0.04
(Ionian Sea)	500	21.47	66±2	0.034±0.008	1.10±0.08	0.23 ±0.05
Date of Coll.:	1000	21.43	21±1	0.016±0.006	0.38±0.03	0.14 ±0.02
12 April 1977	1500	21.41	11±1	0.013±0.003	0.20±0.02	-
Depth: 3490 m	2000	21.40	11±1	0.012±0.008	0.25±0.02	0.13 ±0.03
No. AT-02	S**	21.59	125±6	0.04 ±0.01	1.3 ±0.1	0.072±0.008
Coord.: 32°57'N	50	21.58	166±11	-	1.1 ±0.1	0.16 ±0.03
31°56'E	100	21.61	112±6	0.07 ±0.01	1.25±0.07	0.16 ±0.03
(Levantine Basin)	225	-	83±2	0.06 ±0.01	1.4 ±0.1	0.10 ±0.01
Date of Coll.:	500	21.46	85±4	-	-	0.35 ±0.05
21 April 1977	750	21.46	13±1	-	1.1 ±0.2	0.61 ±0.04
Depth: 1613 m	1000	21.44	8±1	0.012±0.005	0.25±0.03	0.23 ±0.02
	1500	-	3±1	-	0.16±0.02	0.14 ±0.02

*Uncertainties are given in 1σ propagated errors.
**S = surface.

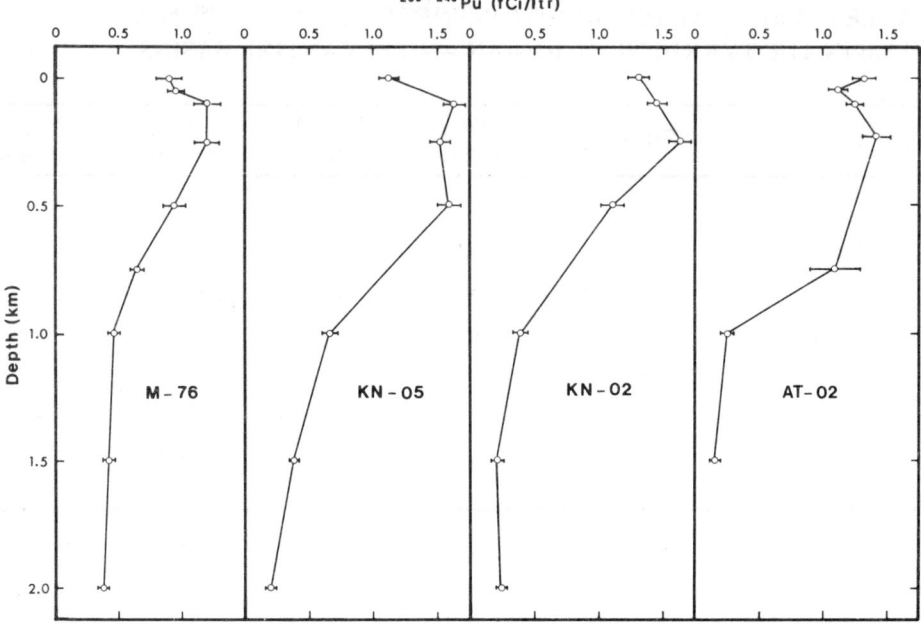

FIG.2.   Vertical distribution of $^{239+240}$Pu at different stations in the Mediterranean.

3.   RESULTS AND DISCUSSION

The results of the measurements of ^{137}Cs, ^{238}Pu, $^{239+240}$Pu and ^{241}Am for
the samples collected from the vertical profiles in the Tyrrhenian Sea (St. KN-05),
the Ionian Sea (St. KN-02) and the Levantine basin (St. AT-02) are presented
in Table I. Based on these data the vertical distribution of $^{239+240}$Pu at each of
the above stations is drawn and compared in Fig. 2 with that in the northwestern
Mediterranean (off Monaco, St. M-76). In each profile shown in Fig. 2 a subsurface
concentration maximum of $^{239+240}$Pu appears near surface around 100–250 m,
followed by a steep decrease between 500 and 1000 m. The less steep decrease
in the mid-depths as well as a relatively homogeneous distribution of $^{239+240}$Pu
below 1000 m for the off-Monaco profile compared with the others are considered
to represent the influence of the more rapid penetration of surface water masses
into depths by the vertical water mixing process characteristic of the northwestern
Mediterranean [6]. For comparison, the vertical distributions of ^{137}Cs in these
profiles are also illustrated in Fig. 2. The general vertical distribution pattern of
^{137}Cs in each profile is similar to the corresponding profile of $^{239+240}$Pu, except
that the concentration maxima of ^{137}Cs tend to appear not only sharper than
those of $^{239+240}$Pu, but also at shallower depths in the Ionian basin (St. KN-02)

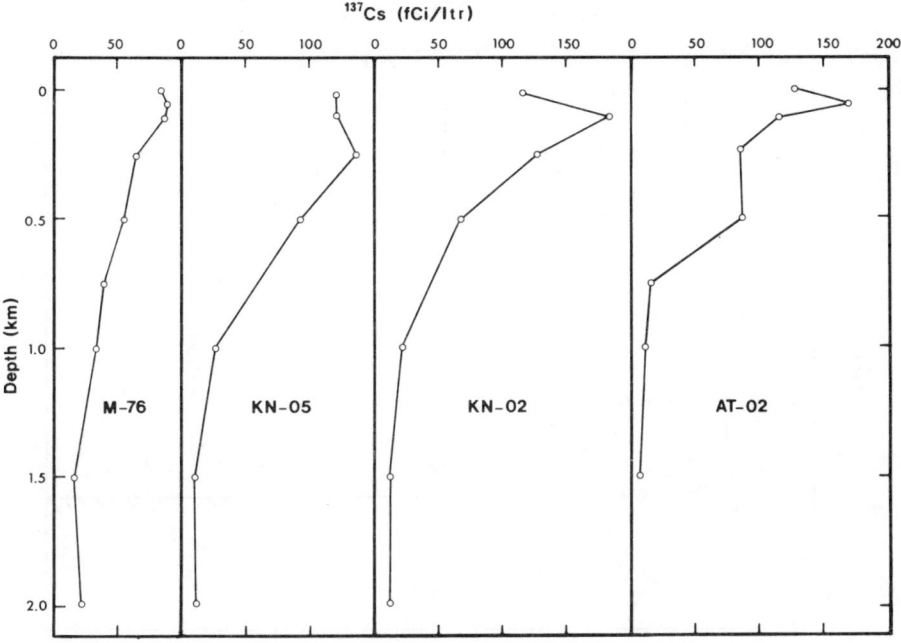

FIG.3.   Vertical distribution of ^{137}Cs at different stations in the Mediterranean.

and the Levantine basin (St. AT-02). As has been pointed out previously for the off-Monaco profile [5], these different vertical distribution patterns between ^{137}Cs and $^{239+240}$Pu indicate that $^{239+240}$Pu is depleted in the near-surface layers relative to ^{137}Cs due probably to its preferential association with sinking particulate matter. Figures 2 and 3 show that this characteristic fractionation between $^{239+240}$Pu and ^{137}Cs in the near-surface layers seems more pronounced in the eastern basins than in the western Mediterranean.

The values of the ^{241}Am/$^{239+240}$Pu activity ratio were computed on the basis of the data given in Table I and the vertical variation of these ratios with depth at Sts. KN-02 and AT-02 is compared with that at St. M-76 in Fig. 4. The figure shows that the characteristic trend of the vertical variation of the ^{241}Am/$^{239+240}$Pu activity ratio observed at St. M-76 [5] holds also at the eastern stations; that is, the activity ratios are lower in near-surface layers (~ 0.1) and higher in deep layers ($\geqslant 0.3$), although the values are much higher below 1000 m at St. AT-02, exceeding 0.8. These are the highest ^{241}Am/$^{239+240}$Pu activity ratios among those observed in Mediterranean waters and close to those reported for Mediterranean sediments (~ 1.0) by Livingston et al. [10]. The cause of these high ratios may be attributed to the abundant occurrence in this area, due to the Nile discharge, of sinking particulate matter, which scavenges effectively ^{241}Am relative to $^{239+240}$Pu

FUKAI et al.

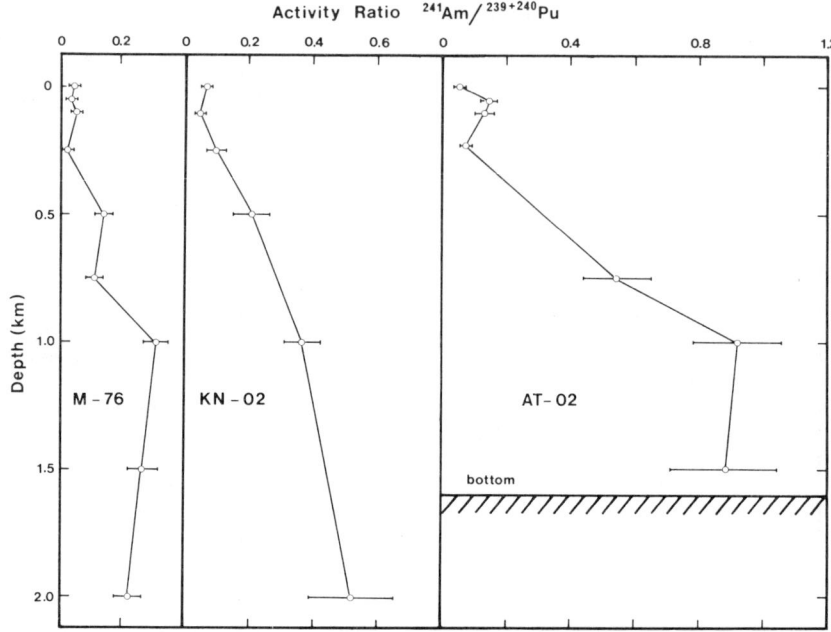

Activity Ratio $^{241}Am/^{239+240}Pu$

FIG.4.   *Variations of the* $^{241}Am/^{239+240}Pu$ *activity ratio with depth at different stations in the Mediterranean.*

in the water column. In fact, at this station, the particulate fractions of $^{239+240}Pu$ and ^{241}Am exceed 30 and 60%, respectively, of the total concentrations of these nuclides in the surface water (unfiltered; unpublished data). These fractions are usually ~ 4 and $\sim 10\%$ respectively for $^{239+240}Pu$ and ^{241}Am in surface waters in other Mediterranean areas studies [11]. Nevertheless, Fig. 4 indicates that the increasing trend of the $^{241}Am/^{239+240}Pu$ activity ratio with depth is widespread in the western Mediterranean as well as the eastern basins. This indicates that the apparent fractionation between $^{239+240}Pu$ and ^{241}Am in the upper layers suggested on the basis of the data from the western Mediterranean is not solely related to the specific vertical water mass transport prevailing in the western Mediterranean, but may be correlated also with mechanisms of vertical material transport specific to the Mediterranean Sea. Thus the fractionation between plutonium and americium specific to the Mediterranean Sea is now considered to characterize the geochemical behaviour of these elements in the Mediterranean as a whole, although further studies may reveal that the same process is taking place in other parts of the oceans.

In Table II the water column inventories of ^{137}Cs, $^{239+240}Pu$ and ^{241}Am computed by integrating the profile data are presented and compared with the

TABLE II. WATER COLUMN INVENTORIES OF ^{137}Cs, $^{239+240}$Pu AND
^{241}Am IN VARIOUS LOCATIONS IN THE MEDITERRANEAN SEA

Station No.	M-76	KN-05	KN-02	AT-02
Mediterranean Basin	North-western	Tyrrhenian	Ionian	Levantine
Date of collection	25 Aug. 1976	15 Apr. 1977	12 Apr. 1977	21 Apr. 1977
Station depth	∿ 2200 m	3577 m	3490 m	1613 m
Fallout delivery				
^{137}Cs (nCi/m^2)	110 + 22	104 + 27	97 + 30	86 + 28
$^{239+240}$Pu (nCi/m^2)	2.1 + 0.5	2.0 + 0.5	1.9 + 0.6	1.6 + 0.5
^{241}Am (nCi/m^2)	0.5 + 0.1	0.4 + 0.1	0.4 + 0.1	0.4 + 0.1
Integrated activity (0 – 2000 m)				
^{137}Cs (nCi/m^2)	77 + 1	102 + 2	97 + 2	67 + 2*
	(70 + 14%)	(98 + 26%)	(100 + 31%)	(78 + 26%)
$^{239+240}$Pu (nCi/m^2)	1.32 + 0.04	1.72 + 0.04	1.33 + 0.04	1.21 + 0.09*
	(63 + 15%)	(86 + 23%)	(70 + 22%)	(76 + 24%)
^{241}Am (nCi/m^2)	0.21 + 0.02	–	0.30 + 0.03	0.41 + 0.02*
	(42 + 9%)	–	(75 + 20%)	(103 + 26%)

* Integrated for 0 – 1500 m.   Note: 1 Ci = 3.70 × 10^{10} Bq.

fallout delivery of these nuclides. The fallout delivery of ^{137}Cs at different latitudes was estimated from the latitudinal distribution of deposited ^{90}Sr in soil up to 1971 [12] by applying the ^{137}Cs/^{90}Sr activity ratio of 1.49 [13]. The delivery of [239+240]Pu was similarly derived from the soil data reported by Hardy et al. [14], while the [241]Am delivery was computed on the basis of the [239+240]Pu delivery by applying the [241]Am/[239+240]Pu activity ratio of 0.22 in integrated global fallout reported for 1974 [15]. The percentages of the integrated activity for the water column (0–2000 m) at different stations against the fallout delivery estimated for the respective latitudes are given in brackets in Table II. Due to the large uncertainties associated with the delivery values, the uncertainties of these percentages are also large. Nevertheless, the percentages for ^{137}Cs and [239+240]Pu show that the major fractions of these nuclides delivered are still in the water column, although the lateral advection of water masses must be taken into account for more precise inventory considerations. Systematic understanding of the [241]Am inventory is much more complicated, since the ingrowth of [241]Am in situ from [241]Pu after the delivery by fallout should be taken into consideration. Although the high percentage of the water column inventory of [241]Am at St. AT-02 is likely to reflect the effects of suspended matter injected from land into sea water by the Nile discharge, further quantitative consideration is not possible at present due to the large uncertainties involved in the delivery terms. Further studies of the vertical transport of transuranic nuclides in Mediterranean water columns that are located especially close to the major river discharges are still required to understand the deposition dynamics of [241]Am.

ACKNOWLEDGEMENTS

The authors wish to express their cordial thanks to the scientific staff and crews of USNS "Kane" and R.V. "Atlantis II" for the assistance rendered during the sampling cruise. Technical assistance rendered by Mrs. D. Vas is also highly appreciated.

REFERENCES

[1] BOWEN, V.T., WONG, K.M., NOSHKIN, V.E., Plutonium-239 in and over the Atlantic Ocean, J. Mar. Res. **29** (1971) 1.

[2] LIVINGSTON, H.D., BOWEN, V.T., "Americium in the marine environment – Relation to plutonium", Environmental Toxicity of Aquatic Radionuclides: Models and Mechanisms, Ann Arbor Sci. Publ., Inc., Ann Arbor, Mich. (1976) 107.

[3] MIYAKE, Y., SUGIMURA, Y., "The plutonium content of Pacific Ocean waters", Transuranium Nuclides in the Environment (Proc. Symp. San Francisco, 1975), IAEA, Vienna (1976) 91.

[4] MURRAY, C.N., FUKAI, R., Measurement of $^{239+240}$Pu in the northwestern Mediterranean, Estuar. Coast. Mar. Sci. **6** (1978) 145.

[5] FUKAI, R., HOLM, E., BALLESTRA, S., A note on vertical distribution of plutonium and americium in the Mediterranean Sea, Oceanol. Acta **2** (1979) 129.

[6] STANLEY, D.J., CITA, M.B., FLEMMING, N.C., KELLING, G., LLOYD, R.M., MILLIMAN, J.D., PIERCE, J.W., RYAN, W.B.F., WEILEY, Y., "Guidelines for future sediment-related research in the Mediterranean Sea", in The Mediterranean Sea: A Natural Sedimentation Laboratory, Dowden, Hutchinson and Ross, Inc., Stroidsburg, Pennsylvania (1972) 723.

[7] BALLESTRA, S., HOLM, E., FUKAI, R., "Low-level determination of transuranic elements in marine environmental samples", in Determination of Radionuclides in Environmental and Biological Materials (Proc. Symp. London, 1978), Central Electricity Generating Board, London (1978) paper No. 15.

[8] HOLM, E., FUKAI, R., A method for ion-exchange separation of low-levels of americium in environmental materials, Talanta **26** (1979) 791.

[9] HOLM, E., PERSSON, R.B.R., "Pu-241 and Am-241 in the environment" (Proc. 4th Int. Congr. IRPA, Paris) **3** (1977) 845.

[10] LIVINGSTON, H.D., CASSO, S.A., BOWEN, V.T., BURKE, J.C., Soluble and Particle-Associated Fallout Radionuclides in Mediterranean Water and Sediments, Rapp. Comm. int. Mer Médit. **25/26** 5 (1979) 71.

[11] HOLM, E., BALLESTRA, S., FUKAI, R., BEASLEY, T.M., Particulate plutonium and americium in Mediterranean surface waters, Oceanol. Acta **3** (1980) 157.

[12] HARDY, E.P., KREY, P.W., VOLCHOK, H.L., Global Inventory and Distribution of Pu-238 from SNAP-9A, HASL Report 250, Health and Safety Laboratory, USAEC, New York (1972) 32pp.

[13] VOLCHOK, H.L., BOWEN, V.T., FOLSOM, T.R., BROECKER, W.S., SCHUERT, E.A., BIEN, G.S., "Oceanic distribution of radionuclides from nuclear explosions", Radioactivity in the Marine Environment, US Natl. Acad. Sci. Washington, D.C. (1971) 42.

[14] HARDY, E.P., KREY, P.W., VOLCHOK, H.L., Global inventory and distribution of fallout plutonium, Nature **241** (1973) 444.

[15] KREY, P.W., HARDY, E.P., PACHUCKY, C., ROURKE, F., COLUZZA, J., BENSON, W.F., "Mass isotopic composition of global fallout plutonium in soil", Transuranium Nuclides in the Environment (Proc. Symp. San Francisco, 1975), IAEA, Vienna (1976) 671.

PLUTONIUM AND OTHER ARTIFICIAL RADIONUCLIDES IN THE SEINE ESTUARY AND ADJACENT AREAS

C. JEANDEL, J.-M. MARTIN, A.J. THOMAS
Laboratoire de géologie,
Ecole normales supérieure,
Paris,
France

Abstract

PLUTONIUM AND OTHER ARTIFICIAL RADIONUCLIDES IN THE SEINE ESTUARY AND ADJACENT AREAS.

Radioactive plutonium and caesium isotopes from the La Hague nuclear fuel reprocessing plant are used as tracers of sediment dynamics in a comparison of three river-estuary systems in France. The low activities of both the dissolved and suspended fractions of these radionuclides compared to natural activity levels in the brackish water are due to enrichment ascribed to the nuclear industrial effluents.

1. INTRODUCTION

The French nuclear fuel reprocessing plant at La Hague discharges low-level radioactive liquid wastes in the southern coastal zone of the Channel. Part of the contaminated waters and sediments may reach the Seine river bay and estuary. Windscale plant (United Kingdom) wastes may also contribute to the radioactivity of this area. La Hague discharges are still poorly documented in the literature but do not reach so far the levels of activity released in the Irish Sea. However, background activity resulting from atmospheric fallout is markedly enhanced by La Hague plant discharges which provide a potential tool for tracing sediment and water dynamics. In particular, radionuclide activity in suspended matter may help understanding of the main origin of sediment supply in the Seine estuary. Data described herein represent a first attempt to evaluate the activity level, fate and distribution of plutonium isotopes in this area, along with the study of the main gamma-emitting artificial radionuclides such as ^{137}Cs, ^{106}Ru, ^{125}Sb and ^{144}Ce.

2. THE AREA IN GENERAL

The area under study is located eastwards of the La Hague nuclear centre, including the coastal zone of the Seine Bay and estuary (Fig.1). Strong tidal

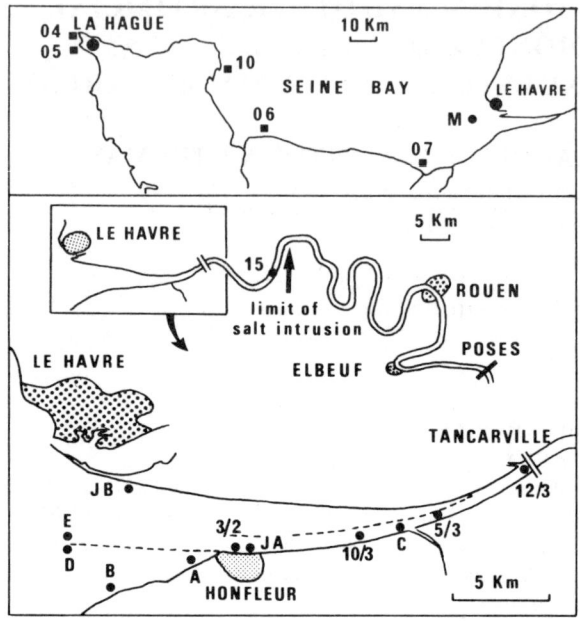

FIG.1. *Sample location: Seine Bay and estuary. Seine river reference station at Elbeuf.*

currents spread contaminated waters and sediments southwestwards along the coasts of Brittany and eastwards to the Seine Bay and even further [1]. Mixing with the river Seine discharge occurs in the estuary and seaward. The main hydrological features of the estuary [2] are briefly summarized below:

Drainage area: 74 250 km²
Main river discharge: 1.5×10^{10} m³/a
Solid discharge: about 1.6×10^6 t/a
Tidal amplitude near mouth: 2 – 8 m
Extension of salt intrusion: $\lesssim$ 50 km from river mouth
Extension of tidal zone: $\leqslant$ 150 km from river mouth (at Poses barrage).

The importance of tides versus river discharge favours water mixing and this estuary belongs to the partially mixed type according to Pritchard's classification, but the estuarine residual circulation has not been clearly described yet. Sedimentary processes are poorly documented. A turbidity maximum (about 1 g · ltr⁻¹) is located in the lower estuary, but the suspended matter's origin and dynamics are still unknown. This suspended matter is characterized by its high carbonate content, reaching 40 to 50% $CaCO_3$ [3].

This property, combined with the high dissolved carbonate concentration in Seine river water may influence trace element distribution. Important oxygen depletion is usually observed in summer in the middle and lower estuaries.

3. SAMPLING AND ANALYTICAL PROCEDURES

Deposited sediments were collected along the shore in the whole area, and oven-dried at 105°C. Water samples of about 200 ltr were collected in the estuary and river (Fig.1). Suspended matter was recovered on the spot by continuous centrifugation at 11 000 rev/min, followed by pressure filtration on 0.4 μm Nuclepore filters. After acidification to pH 1.5 with HCl RP, filtered water was spiked within ^{242}Pu and ^{133}Cs. Dissolved plutonium isotopes and ^{137}Cs were coprecipitated with ferric hydroxides and ammonium molybdophosphate (AMP), respectively. Gamma-emitting radionuclides were analysed by low-level Ge(Li) spectrometry in suspended and deposited sediments, and AMP precipitates. Particulate plutonium was extracted by aqua regia. Leaching solutions and hydroxide precipitates were radiochemically purified with anionic resin Bio-rad AG 1-X8, 50–100 mesh, and plutonium was finally electroplated and analysed by alpha spectrometry with Si-Au surface-barrier detectors. Results of these analyses are reported in Tables I to III, along with Al content in sediments and water chlorinities.

4. DISCUSSION

4.1. General comparison of estuarine sediment activities

In Fig.2 ^{137}Cs and $^{239-240}$Pu activities of deposited and suspended sediments are compared in three different estuarine environments: Seine Bay and estuary, Gironde estuary (only contaminated by atmospheric fallout), and Rhine-Scheldt delta (possibly also contaminated by La Hague and Windscale plant wastes). For both radionuclides the highest activities are observed in the Seine area. Similar relationships have been obtained between ^{137}Cs and other radionuclides (^{106}Ru, ^{125}Sb and ^{144}Ce). Mean activities (Table IV) confirm higher activity levels in the Seine Bay and estuary, the highest differences being observed for ^{106}Ru. Furthermore, on-going investigations have shown that activities in the Rhine-Scheldt deltaic sediments are comparable to those of the Gironde estuary for ^{137}Cs, $^{239-240}$Pu and ^{125}Sb (^{106}Ru and ^{144}Ce activities are below detection limits in this former area). These activities result essentially from atmospheric fallout.

TABLE I. SPECIFIC ACTIVITIES OF ARTIFICIAL RADIONUCLIDES IN SUSPENDED AND DEPOSITED SEDIMENTS OF THE SEINE BAY AND ESTUARY

Sample	Date	Cl (‰)	Al (%)	239-240 Pu (fCi·g⁻¹)	238 Pu (fCi·g⁻¹)	137 Cs (pCi·g⁻¹)	106 Ru (pCi·g⁻¹)	125 Sb (pCi·g⁻¹)	144 Ce (pCi·g⁻¹)
Seine river at Elbeuf (suspended matters)									
S 79.02	Feb.79	0.0	3.06	14 ± 2	—	0.44 ± 0.04	<	<	<
S 79.04	May 79	0.0	1.21	3.6 ± 0.6	—	0.25 ± 0.02	<	<	<
S 79.05	June 79	0.0	0.93	8.7 ± 1.6	—	0.28 ± 0.03	<	<	<
S 79.06	July 79	0.0	1.13	8.1 ± 1.8	—	0.19 ± 0.02	<	<	<
S 79.10	Oct.79	0.0	1.90	6.5 ± 1.2	—	0.48 ± 0.03	<	<	<
S 79.12	Dec.79	0.0	3.81	7.0 ± 1.9	—	0.73 ± 0.09	<	<	<
Seine estuary (suspended matters)									
15/123	Nov.78	0.18	3.25	—	—	0.5 ± 0.2	<	<	—
5/3	Nov.78	0.19	3.59	51 ± 6	16 ± 3	0.80 ± 0.07	2.8 ± 0.9	0.3 ± 0.2	—
3/2	Nov.78	3.4	3.78	65 ± 9	13 ± 3	1.4 ± 0.2	3.2 ± 1.7	0.6 ± 0.3	—
12/3	Nov.78	4.1	4.04	60 ± 7	26 ± 4	1.2 ± 0.1	3.5 ± 1.3	0.6 ± 0.3	—
10/3	Nov.78	11.0	3.61	73 ± 10	36 ± 7	1.4 ± 0.1	3.0 ± 0.9	0.4 ± 0.2	—
S 79 E	March 79	0.8	4.66	103 ± 20	—	1.6 ± 0.1	5.0 ± 0.6	0.6 ± 0.1	0.28 ± 0.11
S 79 C	March 79	1.1	2.89	65 ± 7	19 ± 2	1.2 ± 0.1	3.7 ± 0.4	0.3 ± 0.1	<
S 79 A	March 79	3.0	2.83	53 ± 12	12 ± 5	1.1 ± 0.1	3.4 ± 0.4	0.3 ± 0.1	0.73 ± 0.16
S 79 B	March 79	8.3	1.26	24 ± 3	5.5 ± 1.0	0.37 ± 0.03	1.2 ± 0.2	0.11 ± 0.04	<
S 79 D	March 79	11.6	4.38	99 ± 23	36 ± 19	1.83 ± 0.09	4.5 ± 0.6	0.8 ± 0.2	<
S 79 M	March 79	17.4	3.80	85 ± 13	21 ± 6	1.15 ± 0.07	6.8 ± 0.6	0.4 ± 0.1	0.3 ± 0.1
J A	July 79	2.8	1.08	—	—	0.33 ± 0.04	(0.9)	0.16 ± 0.10	<
J B	July 79	8.3	2.40	43 ± 9	17 ± 8	0.88 ± 0.04	2.8 ± 0.3	0.34 ± 0.06	0.24 ± 0.08
Seine Bay (fine-grained sediments)									
HG.80.07	Jan.80		3.45	59 ± 5	29 ± 3	0.58 ± 0.02	3.5 ± 0.1	0.24 ± 0.02	0.20 ± 0.04
HG.80.06	Jan.80		3.86	78 ± 8	21 ± 2	1.00 ± 0.02	4.1 ± 0.1	0.24 ± 0.02	0.26 ± 0.04
HG.80.10C	Jan.80		3.24	107 ± 4	29 ± 2	0.81 ± 0.02	6.9 ± 0.2	0.35 ± 0.03	0.57 ± 0.06
La Hague area (coarse detrital sands)									
HG.80.04	Jan.80		3.29	6.4 ± 1.2	3.3 ± 0.9	0.07 ± 0.01	0.42 ± 0.03	(0.01)	0.07 ± 0.02
HG.80.05	Jan.80		6.63	76 ± 2	40 ± 2	2.47 ± 0.03	15.3 ± 0.3	2.2 ± 0.1	7.3 ± 0.1

Note: (< = below detection limits).

TABLE II. SPECIFIC ACTIVITIES OF GAMMA-EMITTING ARTIFICIAL RADIONUCLIDES IN SUSPENDED MATTERS OF THE GIRONDE ESTUARY (1979)

Sample	Date	Cl(‰)	Al(%)	^{137}Cs (pCi·g^{-1})	^{106}Ru (pCi·g^{-1})	^{125}Sb (pCi·g^{-1})	^{144}Ce (pCi·g^{-1})
G 79.01	Oct.79	0.0	5.32	0.30 ± 0.05	<	<	<
G 79.10	Oct.79	0.0	8.20	0.48 ± 0.05	0.29 ± 0.08	(0.04)	<
G 79.02	Oct.79	0.1	7.09	0.34 ± 0.02	<	<	<
G 79.31	Oct.79	1.7	8.62	0.43 ± 0.02	0.26 ± 0.08	(0.06)	0.09 ± 0.03
G 79.25	Oct.79	8.3	7.51	0.45 ± 0.1	(0.2)	<	<
G 79.29	Oct.79	17.2	5.98	0.31 ± 0.04	<	(0.14)	<
PK 60	May 79	–	8.10	0.47 ± 0.02	<	<	(0.06)

Note: < = below detection limits.

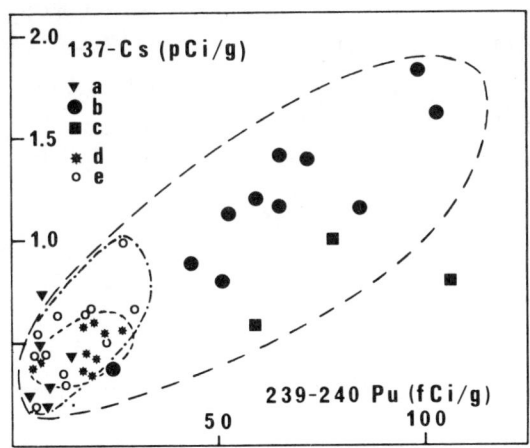

FIG.2. *Caesium − plutonium relationships in various areas:*
a = Seine river suspended matter;
b = Seine estuary (Cl > 0.2‰) suspended matter;
c = Seine Bay fine-grained deposited sediments;
d = Gironde estuary suspended matter (after Ref. [8]);
e = Rhine-Scheldt surface sediments.

TABLE III. SPECIFIC ACTIVITY OF DISSOLVED PLUTONIUM AND
CAESIUM IN THE SEINE ESTUARY (< 0.4 μm except (a) < 0.05 μm)

Sample	Date	Cl (‰)	$^{239-240}$Pu (fCi·ltr^{-1})	^{137}Cs (fCi·ltr^{-1})
Seine river at Elbeuf				
S 79.01	Jan.79	0.0	0.030 ± 0.009	78 ± 7
S 79.02	Feb.79	0.0	0.034 ± 0.010	31 ± 3
S 79.04	May 79	0.0	0.12 ± 0.07	66 ± 7
S 79.05	June 79	0.0	0.23 ± 0.05	44 ± 5
S 79.06	July 79	0.0	0.046 ± 0.017	87 ± 9
S 79.10	Oct.79	0.0	0.13 ± 0.03	80 ± 8
S 79.12	Dec.79	0.0	0.06 ± 0.02	62 ± 6
Seine estuary				
S 79 E	Mar.79	0.8	0.058 ± 0.027	—
			< 0.02 (a)	—
S 79 C	Mar.79	1.1	0.02 ± 0.01	117 ± 12
S 79 A	Mar.79	3.0	0.037 ± 0.012	110 ± 20
S 79 B	Mar.79	8.3	0.078 ± 0.039	360 ± 30
S 79 D	Mar.79	11.6	—	540 ± 50
S 79 M	Mar.79	17.4	0.17 ± 0.08	550 ± 50
J A	July 79	2.8	0.04 ± 0.01	205 ± 35
J B	July 79	8.3	0.09 ± 0.03	506 ± 50
La Hague area				^{238}Pu (fCi·ltr^{-1})
HG 8004		19.7	1.7 ± 0.2	0.7 ± 0.2

TABLE IV. SPECIFIC ACTIVITIES OF ARTIFICIAL RADIONUCLIDES IN SEDIMENTS FROM THE SEINE BAY AND ESTUARY, AND THE GIRONDE ESTUARY

(Mean value and standard deviation, or total range)

Fine-grained deposited sediment		Estuarine suspended matter	
Seine Bay		Seine (Cl > 0.2‰)	Gironde
$^{239-240}$Pu (fCi·g^{-1})	60 − 110	65 ± 24	18 ± 6[a]
^{137}Cs (pCi·g^{-1})	0.6 − 1.0	1.1 ± 0.5	0.46 ± 0.09[a]
			0.41 ± 0.07
^{106}Ru (pCi·g^{-1})	3.5 − 6.9	3.4 ± 1.6	≤ 0.29
^{125}Sb (pCi·g^{-1})	0.24 − 0.35	0.4 ± 0.2	≤ 0.14
^{144}Ce (pCi·g^{-1})	0.2 − 0.6	0.4 ± 0.2	≤ 0.09

[a] After Ref. [8].

4.2. Normalization of activities

The sample matrix must be taken into account in order to allow meaningful comparisons between activities in different sediment samples. It has long been established that trace-element content in fine-grained sediments is generally correlated with clay content, the latter being characterized by aluminium concentration. The following linear correlations have been obtained in the Seine brackish-water suspensions[1] :

$$^{239-240}\text{Pu (fCi·g}^{-1}) = 21 \text{ Al (\%)} - 6.6$$
$$^{137}\text{Cs (pCi·g}^{-1}) = 0.35 \text{ Al (\%)} - 0.0$$
$$^{106}\text{Ru (pCi·g}^{-1}) = 1.1 \text{ Al (\%)} + 0.0$$
$$^{125}\text{Sb (pCi·g}^{-1}) = 0.15 \text{ Al (\%)} - 0.0$$
(^{144}Ce: not sufficient data).

---

[1] 1 Ci = 3.70 × 10^{10} Bq.

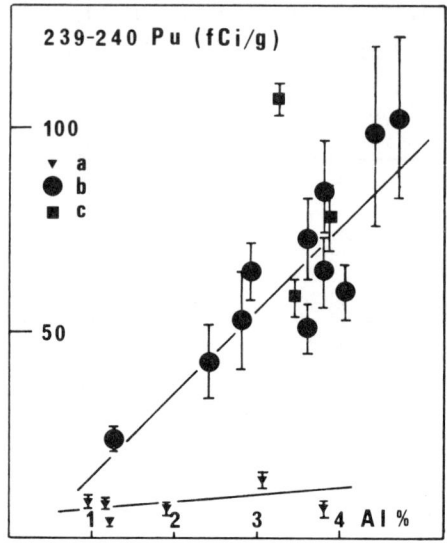

FIG.3. Plutonium — aluminium relationships in the Seine area:
a = river suspended matter;
b = estuarine suspended matter;
c = coastal fine-grained sediments.

However, different relationships with aluminium content may be found in different sedimentary environments (e.g. suspended matter in fresh and brackish water). In the case of pure sand deposits, aluminium content may essentially reflect detrital feldspar distributions. Plutonium/aluminium relationships in fresh- and brackish-water suspensions, and deposited marine coastal sediments, are shown in Fig.3. In view of these linear relationships, and in order to facilitate data interpretation, subsequent sediment activities will be reported as the element-to-aluminium ratio. Figure 4 shows that despite such normalization, activities in suspended matter remain different in the Seine and Gironde estuaries.

4.3. Normalized activities distribution in both estuaries

In fresh-water reference samples, normalized activities of ^{137}Cs and $^{239-240}$Pu are higher in the Seine estuary than in the Gironde estuary (other radionuclides have not been detected). This difference can be ascribed to weathering conditions in the drainage basins, or to a different affinity of sediments for these radionuclides.

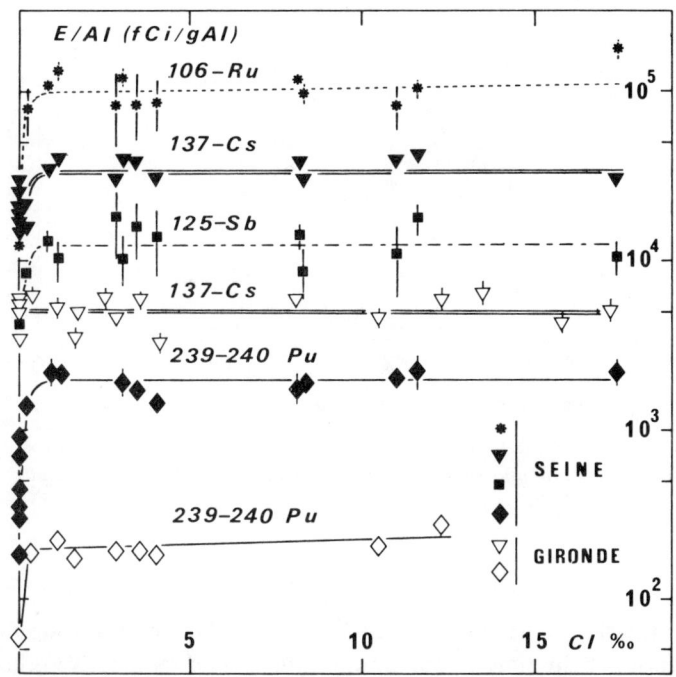

FIG.4. Normalized activities of artificial radionuclides (E) in suspended matters of the Seine and Gironde estuaries, and chlorinity. (Seine River activities for [106]Ru and [125]Sb correspond to detection limits.) Gironde activities after Ref. [8].

In brackish-water samples (Cl > 0.2%) the most prominent feature is over-all constancy of normalized activities for ^{137}Cs and [239-240]Pu in both estuaries. [106]Ru and [125]Sb behave similarly in the Seine estuary but scarcity of data does not allow conclusions to be reached for the Gironde estuary, or for [144]Ce in both cases. This constancy probably means that most of the suspended sediments sampled in the Seine estuary brackish waters have the same origin.

Important activity variations are observed between fresh-water reference samples and brackish-water samples. The case of plutonium will be discussed first. Its activities sharply increase by a factor of 3 in the Gironde and of 4 in the Seine estuary. Several processes may account for this behaviour.

(a) Coagulation in the low-salinity zone of mineral or organic micro-colloids enriched in plutonium can be envisaged: Due to the filtration procedure, such colloids would have been ascribed to the so-called dissolved phase (< 0.4 μm) in fresh-water samples, larger aggregates (> 0.4 μm) due to their coagulation in brackish waters being retained by filters.

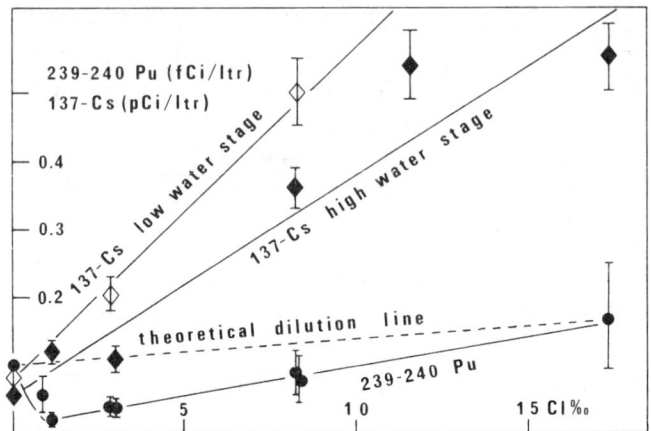

FIG.5.   Dissolved plutonium and caesium activities in the Seine river (mean values) and estuary.

(b)   Sediment particles presenting comparatively higher activities supplied to
      the estuaries at the time of maximum atmospheric radioactive fallout
      (1963). As a result of partial sediment trapping, the mixing of such old
      particles with recent sediment supply may contribute to the plutonium
      increase observed in both estuaries. However, a similar behaviour which
      is observed in the Seine estuary for short-lived radionuclides (0.8 to
      2.8 years half-life) does not support such interpretation. Retention of
      large amounts of 'old' sediments seems unlikely to occur in the Gironde
      estuary where a short residence time of sedimentary particles is generally
      observed [4].

(c)   Actual precipitation of dissolved plutonium should not occur in carbonated
      and slightly alkaline waters of the Gironde and especially the Seine estuary
      owing to its anionic nature in such conditions [5]. Should precipitation
      occur, it must be noticed that dissolved plutonium activities are low
      (< 0.2 fCi $\cdot$ ltr^{-1}) and that removal is not very pronounced (Fig.5).

(d)   This increase may finally result from mixing with more active particles
      originated from the sea. Marine sediment supply to the Gironde estuary
      is quantitatively unimportant [6] but, on the contrary, such an interpretation
      can be envisaged in the Seine estuary.

      In conclusion, coagulation processes may essentially control plutonium
activity increase in suspended matters of estuarine low salinity areas. This
increase may be amplified in the Seine estuary by a supply of enriched particles
originating from the sea.

TABLE V. NORMALIZED ACTIVITIES (10^2 fCi/g Al) OF GAMMA-
EMITTING ARTIFICIAL RADIONUCLIDES IN SUSPENDED MATTERS
OF THE GIRONDE AND SEINE ESTUARIES
(Mean value and standard deviation, or total range)

	Seine river at Elbeuf	Seine estuary (Cl $>$ 0.2‰)	Gironde estuary
^{137}Cs	211 + 58	328 ± 76	51 ± 9[b]
			55 ± 5
^{106}Ru	$\leqslant$ 130[a]	1054 ± 284	$\leqslant$ 30 − 36
^{125}Sb	$\leqslant$ 43[a]	127 ± 34	$\leqslant$ 5 − 23

[a] Estimates.
[b] After Ref. [8].

This latter interpretation is supported by activities and behaviour of other radionuclides (Table V). Caesium-137 normalized activities increase between river reference samples and brackish-water samples by a factor of 1.6 in the Seine estuary, whereas constant activity is observed in the Gironde estuary. Other radionuclides probably show the same behaviour although they were not detected in river suspensions of both estuaries. Activities measured in the Gironde estuary can be considered as representing atmospheric fallout background, and river sample activities in both estuaries are probably rather similar. Then, if we assume on the basis of detection limits that ^{106}Ru and ^{125}Sb activities in the Seine river suspensions are $<$ 0.3 pCi·g^{-1} and $<$ 0.1 pCi·g^{-1} respectively, these activities are multiplied by factors of 8 and 3 in the Seine brackish-water samples. These results confirm the previous hypothesis of introduction into the Seine estuary of enriched particles originated from the sea. These particles are detected up to Tancarville bridge in the middle estuary, showing clearly the existence of a flux of recent suspended sediments from the sea to the upper estuary.

4.4. Origin of contamination

In an attempt to identify the origin of this marine contamination, special attention has been given to the ^{238}Pu/$^{239-240}$Pu activity ratio. ^{238}Pu activities could not be measured in suspended matters of the Seine river, but it seems reasonable to assume that their plutonium activity ratio is close to the atmospheric fallout ratio, i.e. 0.05. Plutonium activity ratios in the Seine area are given below:

Seine estuary suspensions (Cl > 0.2‰): 0.32 ± 0.10
Seine Bay fine-grained sediments: 0.27–0.49
Coarse sands near La Hague plant pipeline: 0.51–0.52
Waters near La Hague plant pipeline: 0.41.

The mean activity ratio of plutonium in 22 deposited sediments (silts and muds) or the Irish Sea near Windscale nuclear centre is 0.24, ranging from 0.2 to 0.3 [7]. Assuming that sediments originated from this area mix with the Seine river suspended material, the resulting activity ratio would be less than 0.24. On the contrary, a higher ratio is found, and the highest values correspond to the La Hague area. It is then most likely that the main origin of contamination in the Seine area is the effluents released by the La Hague plant.

4.5. Dissolved caesium and plutonium

Specific activities of dissolved ^{137}Cs and $^{239-240}Pu$ in the Seine estuary are reported in Table III and Fig.5. Dissolved caesium activities show a fivefold increase from the river to the sea, markedly during the low water period (July 1979). Sea water activities reach 500 fCi·ltr^{-1}. In the Gironde estuary [8], caesium distribution also shows a seaward increase, following approximately the theoretical dilution curve between fresh and sea water, but activities do not exceed 80 fCi·ltr^{-1}. It is, therefore, concluded that atmospheric fallout only cannot account for the high dissolved caesium activities found in the Seine estuary and that sea water in this area is also contaminated by nuclear effluents.

Dissolved $^{239-240}Pu$ activities are of the same order of magnitude in the Gironde estuary [8] and the Seine estuary. A similar removal with respect to the theoretical dilution curve is observed in both cases, and it probably results from coagulation processes previously mentioned. Colloidal plutonium occurrence is confirmed by analyses of sample S 79 E, where 'dissolved' plutonium activity in the filtrate is reduced by a factor of 3 if ultrafiltration (< 0.05 μm) is used instead of filtration (< 0.4 μm).

Activities of these two radionuclides in particulate and dissolved phases can be compared using their distribution coefficients K_d (ratio of specific activity in sediments to specific activity in water). The following values have been obtained in the Seine estuary (Cl > 0.2‰):

K_d (^{137}Cs) = 4.3 × 10^3 ltr·kg^{-1}
K_d ($^{239-240}Pu$) = 1.3 × 10^6 ltr·kg^{-1}

High plutonium K_d emphasizes its strong affinity for the fine particulate phase. As a matter of fact, Seine estuarine water contamination by plutonium is essentially restricted to the suspended matter. Caesium K_d is much lower according to its ionic nature and a significant increase of its dissolved activities is observed in response to water contamination.

5. CONCLUSIONS

Plutonium-239, -240, caesium-137 and other artificial radionuclides in suspended matter of the whole brackish-water zone of the Seine estuary present systematically higher activities than in the Seine river and the Gironde estuary, the latter being considered as representative of atmospheric fallout background. This additional contamination results from introduction into the estuary of enriched particles originating from sea waters. This enrichment is ascribed to industrial effluents, mostly originating from the La Hague nuclear fuel reprocessing plant, as shown by $^{238}Pu/^{239-240}Pu$ activity ratios. Such contaminated particles are used as tracers of sediment dynamics, showing an important sediment flux towards the upper estuary. Resulting activity levels are generally low as compared to the activity of homologous natural radionuclides in suspended matters. Mean activity levels of $^{239-240}Pu$ and ^{137}Cs remain ten times as low as the activity of ^{238}U and ^{40}K, respectively. However, ^{106}Ru presents higher levels, reaching ^{40}K activity.

Plutonium and caesium behaviour was further investigated in the dissolved phase. Dissolved $^{239-240}Pu$ activities are very low and microcolloid coagulation probably accounts for dissolved plutonium removal and contributes to particulate enrichment.

Despite estuarine contamination, dissolved plutonium activity remains ten thousand times lower than the dissolved ^{238}U activity of sea water, according to its strong affinity for the fine particulate phase. On the other hand, in agreement with its ionic nature and a lower affinity for particulate phase, dissolved caesium activity is clearly affected by water contamination though it remains a thousand times lower than the dissolved ^{40}K activity of sea water.

ACKNOWLEDGEMENTS

Dr. J.M. Harley from HASL, United States Atomic Energy Commission, is gratefully thanked for providing a ^{242}Pu spike.

REFERENCES

[1] GUEGUENIAT, P., AUFFRET, J.-P., BARON, Y., Evolution de la radioactivité
     artificielle gamma dans des sédiments littoraux de la Manche pendant les années
     1976–1977–1978, Oceanol. Acta 2 2 (1979) 165.
[2] S.A.U.M., Estuaire de la Seine, Dossier n°1 (1977) 175.
[3] GERMANEAU, J., Etude de la sédimentation dans l'estuaire de la Seine. 2ème partie:
     origine, déplacement et dépôt des suspensions, Trav. Centre Rech. Etudes Océanogr.
     Paris 9 1/4 (1969) 100.
[4] CASTAING, P., JOUANNEAU, J.-M., Temps de résidence des eaux et des suspensions
     dans l'estuaire de la Gironde, J. Rech. Océanogr. 4 2 (1979) 41.
[5] EDGINGTON, D.N., "Characterization of transuranic elements at environmental levels",
     Techniques for Identifying Transuranic Speciation in Aquatic Environments (Proc.
     Joint CEC/IAEA Tech. Meeting Ispra, 1980) these Proceedings.
[6] ALLEN, G.P., SAUZAY, G., CASTAING, P., JOUANNEAU, J.-M., "Transport and
     deposition of suspended sediments in the Gironde estuary, France", Estuarine Processes
     (WILEY, M., Ed.) 2, Academic Press (1976) 63.
[7] HUNT, G.J., Radioactivity in surface and coastal waters of the British Isles, 1977,
     Aquatic Env. Monitoring Rept No.3, Minist. Agr. Fish & Food, Dir. Fish Res.,
     Lowestoft (1976) 36.
[8] CAUWET, G., ELBAZ, F., JEANDEL, C., JOUANNEAU, J.-M., LAPAQUELLERIE, Y.,
     MARTIN, J.-M., THOMAS, A.J., Comportement géochimique des éléments stables et
     radioactifs dans l'estuaire de la Gironde en période de crue, Bull. Inst. Géol. Bassin
     d'Aquitaine, Bordeaux 27 (1980) 5.

EFFECT OF A LONG-TERM RELEASE
OF PLUTONIUM AND AMERICIUM INTO AN
ESTUARINE AND COASTAL SEA ECOSYSTEM

C.N. MURRAY, A. AVOGADRO
Commission of the European Communities,
Joint Research Centre,
Ispra Establishment,
Chemistry Division,
Ispra, Varese,
Italy

Abstract

EFFECT OF A LONG-TERM RELEASE OF PLUTONIUM AND AMERICIUM INTO AN
ESTUARINE AND COASTAL SEA ECOSYSTEM.

This paper discusses the general problem of speciation of plutonium and americium in
aquatic ecosystems and the implications relative to their fate in those systems. The following
conclusions were reached: several oxidation states of plutonium coexist in the natural environment; the effect of environmental changes such as pH and E_h values and complexes are probably
the cause of these various oxidation states; a clearer definition of the 'concentration factor'
should be given in view of the important role the sediments play in supplying plutonium for
transfer through the food web.

1. INTRODUCTION

A methodology for the assessment of the distribution, fate and associated
hazard due to a release of long-lived alpha-emitting radionuclides into surface
water of an estuarine-coastal sea ecosystem is being developed. The study is part
of a project on the assessment of risk associated with the long-term storage of
solidified radioactive wastes in geological formations [1].

The present paper develops the discussion of the problem of chemical
speciation of plutonium and americium and their reactions in an aquatic ecosystem
which was presented in a preliminary article by Murray and Avogadro [2]. The
initial methodology developed by Murray and Avogadro [3] concerned the assessment of the effect of the introduction of plutonium and americium into aquatic
systems. Due to the lack of data it was simply assumed that a single species of
plutonium and americium (IV) and (III), respectively, existed and that their
behaviour was similar. However, recent, environmental data reported by
Bondietti and Reynolds [4], Hetherington and Harvey [5], Murray et al. [6] and
Nelson and Lovett [7] have clearly shown this is not the case; the present analysis
is thus developed to consider these new data, especially in relation to the differences

103

in environmental behaviour of various species of these elements. In a subsequent article by Murray and Avogadro [8] the aspects concerning concentration factors and critical pathway analysis are discussed in detail.

2.   CHEMICAL SPECIATION

The following reactions for the formation of different species of plutonium in the environment can occur

$$PuO_2 + 2 H_2O \rightleftarrows Pu^{4+} + 4 OH^-$$

$$Pu^{4+} + H_2O \rightleftarrows Pu(OH)^{3+} + H^+$$

$$Pu^{4+} + 4 H_2O \rightleftarrows Pu(OH)_4^0 + 4 H^+$$

$$Pu^{4+} \rightleftarrows Pu^{3+} - e^-$$

$$Pu^{4+} + 2 H_2O \rightleftarrows PuO_2^+ + 4 H^+ + e^-$$

$$Pu^{4+} + 2 H_2O \rightleftarrows PuO_2^{2+} + 4 H^+ + 2 e^-$$

$$Pu^{4+} + 3 H_2O \rightleftarrows PuO_2(OH)^+ + 5 H^+ + 2 e^-$$

$$Pu^{4+} + PuO_2^+ \rightleftarrows Pu^{3+} + PuO_2^{2+} \qquad K_{eq} = 13.2$$

Thus, in natural waters, plutonium may exist in different oxidation states (Bondietti and Sweeton [9] and Andelman and Rozzell [10]). Valence state (IV), the most stable under a large range of environmental conditions, is extremely insoluble and exhibits a very strong tendency towards polymer formation; valence states (III), (V) and (VI) are generally more soluble and do not polymerize. Valence state (V), which prevSiously has been considered unstable, may in reality be the species which governs the total plutonium solubility, at low plutonium concentrations and in the absence of a complexing agent. Complexes with carbonate ions are generally considered to be the most stable among inorganic complexes of plutonium, so that when carbonates are present, complex formation with other anions often become negligible. As pointed out by Saltelli et al. [11], substantial data are not yet available or are not reliable for plutonium.

In view of the evident importance of information on the chemical species of actinides in natural waters, the development of techniques for the separation and characterization of chemical species has been undertaken as a first stage towards gaining a better understanding of their behaviour in the environment.

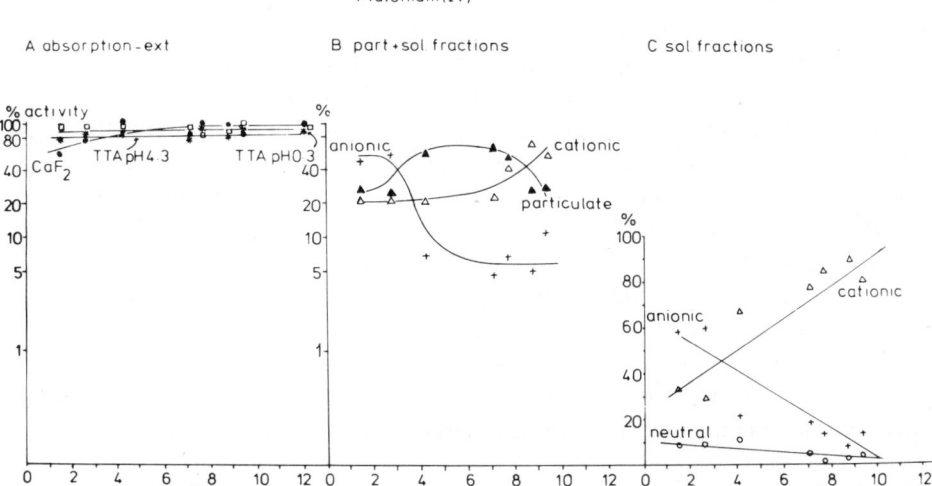

FIG.1.   Results obtained using Pu(IV) chemical speciation in Lake Maggiore water.

In view of this need, methods are thus being developed at the Joint Research Centre, Ispra, to characterize the oxidation forms of the actinides; initial work has been undertaken on plutonium and americium.

The technique employed is based on the work of Foti and Freiling [12] and uses a combination of solvent extraction by thenoyltrifluoroacetone (TTA) and column separation by calcium fluoride and organic resins. The methods have been tested using water samples (freshwater and sea water) contaminated with [237]Pu and [241]Am, where initially known oxidation states have been produced (JRC preliminary report 1979). The pH values of the samples were adjusted to give a range of values that can be found in the environment.

In order to compare the chemical oxidation states determined by TTA extraction, crystalline calcium fluoride is used. The distribution of particulate and soluble species of the various valence states are further determined using an analysis train which gives information on particle size, cationic, anionic and the neutral fractions.

To illustrate the results obtained, Fig. 1 shows data for samples of Lake Maggiore water contaminated over one month with [237]Pu(IV). It can be seen that there is a formation of a particulate fraction (0.45 μm and 0.01 μm) (Fig. 1 B), as well as anionic and cationic species, which vary strongly with the pH value. Taking only the soluble fractions (those which pass through a 0.01μm filter) there appears to be an exchange between the anionic and cationic fractions as the pH value increases. A small neutral fraction can also be seen.

As has been mentioned, complexes of carbonate probably play an important role in allowing the formation of mobile forms of plutonium and americium; the results from lake Maggiore water show the formation of soluble species. Thus, using the techniques described above, studies have been undertaken on the formation of carbonate complexes in order to obtain data on their stability. These data are clearly needed for a better understanding of their kinetic and thermodynamic behaviour at environmental levels. Data already obtained (to be published elsewhere) have shown that these complexes appear to be very sensitive to changes in the pH $-$ E_h values and total carbonate concentration and that both anionic and cationic complexes seem to be formed.

The fact that such large variations in chemical speciation occur will clearly play an important role in a number of important environmental processes and it is thus necessary to identify and develop more realistic models of the controlling mechanisms; the main areas to be considered in this article are the following.

2.1. Sorption and transport

It was shown very clearly by Hetherington [13], Hetherington et al. [14] as well as by others, that sorption processes to suspended material and bottom sediments account for about 95% of the total input of actinides in the Irish Sea, at least over the short term. A differential between the behaviour of plutonium and americium has, however, been reported by Hetherington and Harvey [5] who give results for the distribution of these elements between suspended material and solution in the Irish Sea surface water. These workers report that at a distance of less than 10 km from the point of discharge, the percentage fraction of activity associated with suspended material for plutonium and americium was 36 ± 8 and 80 ± 7, respectively; twice as much americium being associated with suspended material as plutonium. Even at a distance of about 100 km this fraction is still more than 50%.

These results can be explained on the basis of the results of Nelson and Lovett [7] who have shown that two oxidation states of plutonium exist in these waters. Further it appears that the more mobile form is that of Pu(VI), which is in good agreement with thermodynamic data discussed by Aston et al. [15]. Supporting data to the fact that more than one species of plutonium exists can be obtained from Murray et al. [6] who have further shown that a migrating species of plutonium, originating from Windscale, can be detected in the central North Sea and adjacent Atlantic areas.

2.2. Bottom sediment migration

As bottom sediments appear to act, at least in the short term, as the major sink of actinides in coastal environments, migration of different chemical species through the sediment-water interface is of particular importance. This adsorption results in an equilibrium between plutonium in solid and liquid phases. The sediment, over the long term, however, may be able to act as a source for soluble species of plutonium through a series of thermodynamic equilibriums of the following type,

$$K_d = \frac{(Pu^{4+})_{sediment}}{(Pu^{4+})_{water}}$$

$$(Pu^{4+})_{water} \quad \begin{array}{c} \nearrow \quad Pu(IV) \text{ complexes} \\ \searrow \\ PuO_2^{2+} \rightarrow Pu(VI) \text{ complexes} \end{array}$$

resulting in a continuous release of very small quantities of soluble plutonium.

In studies carried out at the JRC Ispra on the interaction of actinides with the environment, preliminary data on the migration of transuranic nuclides through sandy clay columns have demonstrated that several species for plutonium, americium and neptunium occur and that their interaction with the underground formation produces a complex soil migration pattern which is not explainable in terms of a single equilibrium distribution coefficient. In this study it has been found that negatively charged species may be formed which migrate without interaction through soil and sediment. In fact, this is what appears to happen in our column experiments; when the contaminated sub-soils are eluted with uncontaminated ground water there is always a continuous release of about 2×10^{-13} M activity from the column, not explainable by a chromatographic equilibrium process. These soluble species can be interpreted by the formation of anionic complex carbonates not retained by the soil.

Carbonate complexes have also been shown to be the most stable soluble form, under environmental conditions, by Polzer [16] and Bondietti and Sweeton [9]. These complexes are mainly anionic and, therefore, interactions with sediment and suspended particulate materials seem to be small due to the cationic exchange properties of these materials.

2.3. Biological transfer

The factors affecting the accumulation of actinides and their transfer through food chains are not well known. Studies by Fowler and Guary [17] of a simple marine food pathway and by Holm and Persoon [18] of a terrestrial one indicate that the biokinetics of different chemical forms of these elements are extremely complex.

TABLE I. DISTRIBUTION OF PLUTONIUM AND AMERICIUM IN ESTUARY CONDITIONS (UNDER CONDITIONS OF UNIFORM MIXING)

Estuary pH 8.0	Distribution			Activity (pCi·kg⁻¹)		
	Pu(IV)	Pu(VI)	Am(III)	Pu(IV)	Pu(VI)	Am(III)
Total ground water activity (pCi·kg⁻¹)	99	1	100	29.7	0.30	25
Absorbed fraction on suspended freshwater sediment (pH 6.5)	80	95	20	24	0.285	5
Short-term desorption from suspended matter	5	10	10	1.2	0.029	0.5
Original non-absorbed activity in river	20	5	80	6	0.015	20
Total non-absorbed activity in estuary (pH 8.0)	24	14.5	82	7.2	0.044	20.5
Fraction of total having particle size > 0.45 µm	96	32	99	6.9	0.014	20.3
Fraction of total considered 'soluble' < 0.45 µm in estuary water	4	68	1	0.3	0.03	0.2

In the assessment of the behaviour of actinides in the environment, biological transfer, identified using critical pathway analysis, has been shown to play an important role in the return of radioactivity to man and the dose he receives.

At the JRC Ispra an attempt is being made to develop experimental procedures which will allow an insight into certain biological mechanisms identified as being of importance in the development of the long-term assessment of actinides. In this study Thiels et al. [19] are considering chemical speciation of these radio-elements in relation to their uptake and retention for selected biota. Preliminary investigations of their incorporation into cellular material of *L. stagnalis* as well as of the identification of the environmental factors that may have a controlling role in these processes were carried out.

It is clear that the biological incorporation through different uptake routes and the subsequent transfer along food chains are important in the study of biogeochemical cycling of these elements. Fowler [20] has discussed the vertical transport of trace elements through biological processes in the marine environment and pointed out the need for a better understanding of such mechanisms.

3.   IMPLICATIONS FOR THE DEVELOPMENT OF ASSESSMENT METHODOLOGIES

3.1. Plutonium

As an example of the way in which the introduction of the concept of different chemical oxidation states will affect model calculation on the distribution of plutonium in coastal waters — the data assumed by Murray and Avogadro [3] having been considered — the model compartments are shown in Fig. 2.

Table I shows the distribution of plutonium and americium in river and estuary water. It can be seen that the differences in distribution values (K_d) for Pu(IV) and (VI) in freshwater result, in a change in their distribution in the estuarine environment. In the present analysis the initial ratio for total plutonium in ground water was assumed to be 1% (VI) and 99% (IV). The distribution in river water of non-absorbed plutonium is calculated, therefore, to be 0.3% (VI) and 99.7% (IV). In the estuary the 'soluble' fraction distribution is found to be 10% (VI) and 90% (IV). This alteration in the distribution is believed to be due to physico-chemical reactions during the transition between river and sea waters.

Differences in chemical forms of actinides may also explain the results found in Irish Sea water and sediments reported by Hetherington et al. [14]. The interesting point about the sediment water activities shown in Fig. 3 is that they are not parallel and, thus, apparently the global K_d between the sediment and water is not constant. These results can be explained, however, if it is assumed that

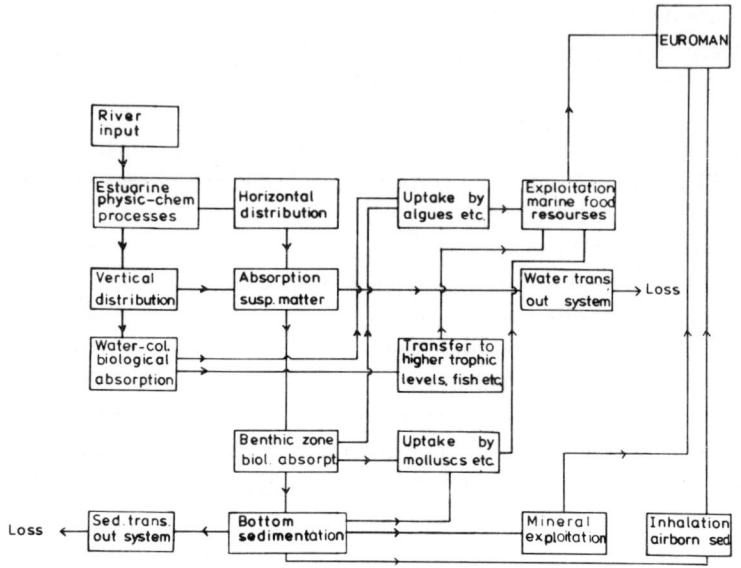

FIG.2.   Model compartments.

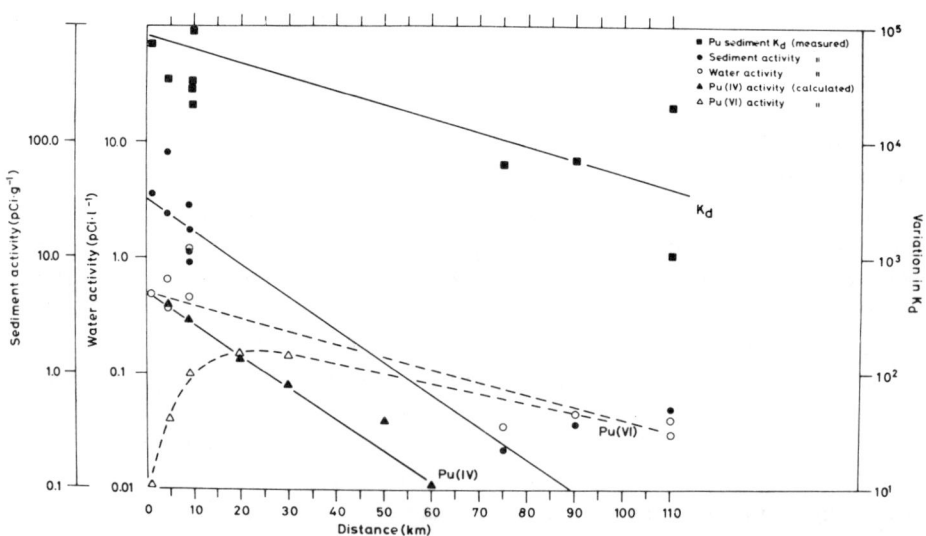

FIG.3.   Distribution of Pu(IV) and (VI) activities with distance in Irish Sea water and sediment.

two oxidation states of plutonium are present in the system. The water concentration of plutonium may be divided into two components (IV) and (VI). The (IV) component is assumed to parallel the sediment activity (i.e. a constant K_d); the concentration for (VI) may be calculated as being the difference between the measured water activities and that calculated for (IV) on the basis of constant K_d (IV). Table II shows the relative activities calculated for the two species in relation to the distance for the input source.

Figure 3 shows the steady decrease of K_d with distance, as the Pu(VI) form becomes the dominant soluble species. On the other hand, the distribution coefficient (K_d) of ^{137}Cs with distance remains quite constant (for mud-type sediments) according to the data of Preston et al. [21], indicating that equilibrium conditions are attained between sea water and sediment over the distances considered in the present example. The fact that two forms of plutonium exist in sea water has been demonstrated by Nelson and Lovett [7] who have also shown that, as indicated in the present analysis, the distribution of the two forms changes with distance, indicating a differential behaviour of the species.

The in situ data show that Pu(V + VI) is more mobile than Pu(III + IV) and will thus cause a non-equilibrium state to be set up between the species. This will have the tendency to cause a constant production of the more mobile species of plutonium. The rate of the kinetic reaction under environmental conditions is not known at present. Nevertheless it seems probable that only the more mobile species (V + VI) is able to move over long distances and that probably the (III + IV) form rests near the input source.

3.2. Americium

The behaviour of other actinide elements such as americium and curium has now been clearly shown to be different from that of plutonium by Dahlman et al. [22]. Data presented by Murray et al. [6] show that americium is much less mobile than plutonium on the basis of the Am/Pu ratio in waters of the Minch. These authors also showed that in waters entering the northern North Sea, the alpha Am/Pu ratio was reduced to about 0.06 from an average of 1.7 in waste discharged at Windscale.

From calculations of K_d based on the formation particulate material retained on a 0.45 μm, Murray and Fukai [23] found a value of between $(0.5 - 1.0) \times 10^5$ in sea water for americium at pH 8.0. This places the americium K_d between those of Pu(VI) of about 2×10^4 and Pu(IV) of greater than 10^6, as reported by Nelson and Lovett [7] for sea water.

The difference in americium behaviour to that of plutonium can clearly be distinguished when the source of the isotopes is the same. However, americium can also be formed by the decay of ^{241}Pu and the chemical form of the ^{241}Pu may

TABLE II. VARIATION OF PLUTONIUM (IV) AND (VI) ACTIVITIES WITH DISTANCE IN IRISH SEA WATER AND SEDIMENT

	1 km	4.5 km	9 km	50 km	90 km
Pu(IV) $\Big\} (pCi \cdot kg^{-1})$	0.49	0.39	0.30	0.03	—
Pu(VI)	0.01	0.04	0.09	0.11	0.05
total	0.50	0.43	0.39	0.14	0.05
Pu $\dfrac{(VI)}{(IV)}$ (%)	2	9	23	79	98
Sediment (pCi·g^{-1})	34	24	18	12	0.1

be either in (IV) or (VI) form. In the case where a significant fraction of [241]Pu is in the (VI) form and therefore more mobile, the distribution of the daughter [241]Am would be very different from [241]Am introduced directly into the environment at the same time as [241]Pu. The americium from this latter source would be immobilized very rapidly in the vicinity of the input.

4.   CONCLUSIONS

The present paper discusses the implications of actinide chemical speciation on their distribution and fate in the aquatic environment.

The major points may be summarized as follows:

(a)   The coexistence of several oxidation states of plutonium in the natural environment is discussed. However, the rates of formation of different species in the aquatic system are unknown. The in situ production may constitute a long-term source of more mobile forms in the environment.

(b)   The effect of the pH and E_h values and complexation reactions is probably the cause of the great variability of actinide species and, thus, will play an important role on their subsequent distribution.

(c)   The effect of different input sources, direct, fallout, radioactive decay etc. on bioavailability must be considered when assessing the probable long-term effects.

(d)   A clearer definition of the concentration factor is required, especially in view of the evident role of sediments as a sink-source term for actinides and the importance of these factors in calculations of dose to man from critical pathway analysis.

REFERENCES

[1] GIRARDI, F., BERTOZZI, G., D'ALESSANDRO, M., Geological Disposal of Radioactive Waste; A Model for Risk Assessment, Commission of the European Communities JRC Ispra Rep. EUR-5902e (1977).
[2] MURRAY, C.N., AVOGADRO, A., A preliminary report on the effect of a long-term release of plutonium and americium into an estuarine-coastal sea ecosystem. II. Chemical speciation and environmental factors, Rapp. Comm. Int. Mar. Médit. 25/26 (1979) 89.
[3] MURRAY, C.N., AVOGADRO, A., Effect of a long-term release of plutonium and americium into an estuarine-coastal sea ecosystem. I. Development of an assessment methodology, Health Phys. 36 (1979) 573.
[4] BONDIETTI, E.A., REYNOLDS, S.A., Field and Laboratory Observations on Plutonium Oxidation States, Oak Ridge National Laboratory Rep. 2117 (1976) 505.

[5] HETHERINGTON, J.A., HARVEY, B.R., Uptake of radioactivity by marine sediments and implication for monitoring metal pollutants, Mar. Pollut. Bull. **9** (1978) 102.

[6] MURRAY, C.N., KAUTSKY, H., HOPPENHEIT, M., DOMIAN, M., Actinide activities in water entering the northern North Sea, Nature **176** (1978) 225.

[7] NELSON, D.M., LOVETT, M.B., The oxidation states of plutonium in the Irish Sea, Nature **276** (1978) 599.

[8] MURRAY, C.N., AVOGADRO, A., Effect of a long-term release of plutonium and americium into an estuarine-coastal sea ecosystem. III. Concentration factors and critical pathway analysis, Health Phys. (1980) (submitted).

[9] BONDIETTI, E.A., SWEETON, F.H., "Transuranic speciation in the environment", Proc. Symp. Transuranics in Terrestrial and Aquatic Environments NVO 179 (1977).

[10] ANDELMAN, J.B., ROZZELL, T.C., Plutonium in the water environment, in Radionuclides in the Environment, Adv. Chem. Ser. 93, Am. Chem. Soc. (1970) 118.

[11] SALTELLI, A., AVOGADRO, A., BERTOZZI, G., "An assessment of plutonium chemical forms in groundwater", NEA/CEC Workshop on Migration of Long-Lived Radionuclides in the Geosphere (1979).

[12] FOTI, S.C., FREILING, E.C., The determination of the oxidation states of tracer uranium, neptunium and plutonium in aqueous media, Talanta **2** (1964) 385.

[13] HETHERINGTON, J.A., The uptake of plutonium nuclides by marine sediments, Mar. Sci. Commun. **4** (1978) 237.

[14] HETHERINGTON, J.A., JEFFERIES, D.F., LOVETT, M.B., "Some investigations into the behaviour of plutonium in the marine environment", Impacts of Nuclear Releases into the Aquatic Environment (Proc. Symp. Otaniemi, 1975), IAEA, Vienna (1975) 193.

[15] ASTON, S.R., AVOGADRO, A., MURRAY, C.N., STANNERS, D.S., "Theoretical and practical problems in the evaluation of physico-chemical forms of transuranics in the marine environments", Impacts of Radionuclide Releases into the Marine Environment (Proc. Symp. Vienna, 1980), IAEA, Vienna (1981) 143.

[16] POLZER, W.L., "Solubility of plutonium in soil/water environments", Safety in Plutonium Handling Facilities (Proc. Symp. Rocky Flats), USAEC Conf. 71 0401 (1971) 411.

[17] FOWLER, S.W., GUARY, J.C., High absorption efficiency for ingested plutonium in crabs, Nature **266** (1977) 827.

[18] HOLM, E., PERSOON, B., Lund University Rep. L-UFD6/(NFRA – 3013) (1977) 1.

[19] THIELS, L., MURRAY, C.N., RADE, J., "Chemical speciation and bioavailability of transuranics for a freshwater snail (*Lymnaea stagnalis* L.)", these Proceedings.

[20] FOWLER, S.W., Trace elements in zooplankton particulate products, Nature **269** (1977) 51.

[21] PRESTON, A., JEFFERIES, D.F., MITCHELL, N.T., "The impact of [134]Cs, ^{137}Cs on the marine environment from Windscale", in Seminar on Radioactive Effluents from Nuclear Fuel Reprocessing Plants, CEC (1977) 401.

[22] DAHLMAN, R.C., BONDIETTI, E.A., EYMAN, L.D., "Biological pathways and chemical behaviour of plutonium and other actinides in the environment", in Actinides in the Environment (1977) 47.

[23] MURRAY, C.N., FUKAI, R., "Adsorption-desorption characteristics of plutonium and americium with sediment particles in the estuarine environment: Studies using [237]Pu and [241]Am", Impacts of Nuclear Releases into the Aquatic Environment (Proc. Symp. Otaniemi, 1975), IAEA, Vienna (1975) 179.

SUMMARY AND RECOMMENDATIONS

1. SUMMARY

It is clear from the papers presented at this meeting that there are many laboratories which now have the capability to measure transuranic elements at environmental levels. However, it is necessary that this resource is now used to the fullest extent to further an understanding of the behaviour of transuranic elements in a variety of environments with widely differing chemical, physical and biological properties.

In discussions following the formal presentation, the group agreed that the state-of-the-art in instrumentation for alpha spectrometry using solid-state detectors is more than adequate for our experiments. Therefore, more attention was paid to problems associated with sample preparation, preconcentration, addition of yield monitors and chemical separations, where errors can arise due to poor technique. Such errors can arise from (i) incomplete chemical separation of different nuclides with similar alpha energies, e.g. ^{238}Pu-^{228}Th or ^{237}Np-^{234}U ^{210}Po-^{243}Am, (ii) inadequate care in pretreatment of water samples prior to the coprecipitation step, resulting in incomplete exchange between the radiochemical yield determinant (spike) and the actinides in the sample or incomplete conversion between oxidation states, and (iii) incomplete separations from lanthanides and iron giving thick deposits resulting in poor resolution alpha spectra.

The group recognized the problems associated with the determination of extremely low concentrations of the transplutonium actinide elements and they felt that more attention should be paid to developing more precise and less time-consuming procedures as well as the provision of suitable samples for inter-comparison. The possibility of group and sequential separation is important in respect of saving time in analysis.

The group accepted the general conclusions presented in the plenary lecture regarding the chemical state of transuranic elements (particularly plutonium) in the environment. That is to say, at environmental concentrations of 10^{-17} to 10^{-13} M plutonium appears to behave as a simple inorganic anion in oligotrophic waters such as the oceans, and at these concentrations is not present as a colloid. However, it is clear that there is still a great deal to be learnt regarding the chemical speciation of plutonium in natural waters, such as ground water and small entrophic lakes, and that a considerable effort must be placed in understanding the basic thermodynamics of the redox reactions for plutonium ions and complex formation with common ligands in the environment as well as the kinetics of interconversion of oxidation states. While analogues such as the natural actinides may be useful in making initial calculations of behaviour, the effective modelling of the behaviour of the transuranic elements is strongly dependent on the availability of these accurate basic thermodynamic constants.

2.   SPECIFIC RECOMMENDATIONS

(1)   In order to expand our knowledge of the behaviour of the actinide elements, studies should be initiated on unique environments which are either already well characterized or have received actinides from sources which have unusual characteristics. Such environments include: the Baltic Sea and the Black Sea, where there are permanent anoxic basins; Thule, where there is an atypical source term; Trombay, where there are both atypical water characteristics and an atypical source term; and freshwater lakes with widely differing limnological characteristics.

(2)   In order to predict the behaviour of the transuranic elements released to the environment either from deep sea dumps or from geological emplacement, fundamental studies of the thermodynamics and kinetics of the interconversion of oxidation states and the stability constants of complexes with ligands of environmental concern at temperatures and pressures related to problems in surface waters, ground waters and the deep sea should be performed.

(3)   The IAEA Reference Manuals for the measurement of radionuclides in marine waters should be updated with particular emphasis on methods for the determination of transuranic and natural alpha emitters as well as techniques for determining oxidation states.

(4)   A supply and distribution of the plutonium and americium isotopes for use as isotopic diluents or yield monitors in the analysis of environmental samples should be established by appropriate international organizations.

(5)   Steps should be taken by appropriate organizations to distribute a range of intercomparison samples for the transplutonium elements of such composition that all workers funded by these organizations will be encouraged to participate in such exercises.

(6)   The chemistry of ground water, particularly for critical ligands which complex actinides, should be established in order to facilitate studies related to the disposal of radioactive wastes in terrestrial environments including geological formations.

(7)   New and improved methods for measuring ultra-low concentrations of actinide elements in environmental samples should be developed.

Session II

METHODS OF STUDYING
THE BIOAVAILABILITY OF TRANSURANICS
IN AQUATIC ORGANISMS

Scientific Secretary

C. MYTTENAERE
CEC, Brussels

Chairmen

O. VANDERBORGHT
Belgium

J. PENTREATH
United Kingdom

RADIOECOLOGICAL SIGNIFICANCE OF TRANSURANICS FOR AQUATIC ORGANISMS
Trends for future research

O. VANDERBORGHT
SCK/CEN,
Mol,
and
University of Antwerp,
Antwerp,
Belgium

Abstract

RADIOECOLOGICAL SIGNIFICANCE OF TRANSURANICS FOR AQUATIC ORGANISMS.
TRENDS FOR FUTURE RESEARCH.
This review paper on the biological availability of transuranics emphasizes where co-operative research by scientists in other related disciplines will provide more insight into the mechanisms that cause biological fixation of transuranics. This information is needed to realistically assess the safety measures that have to be taken to control releases of these radionuclides into the biosphere.

1. INTRODUCTION

This review aims at putting the problem of the biological availability of transuranics to aquatic organisms in the more general context of the continuing problems in radioecological research. This approach has two advantages: (i) it helps to solve problems that are open in other fields of radioecology and (ii) it may avoid a repetition of research already made in other areas of radioecology. This paper is thus not a general review of what is known about biological availability of transuranics, but emphasizes those points where co-operative research by biologists, chemists, hydrologists, microbiologists, et al. will give more insight into the mechanisms underlying the biological fixation of transuranics and thus the safety measures that have to be taken for the actual or potential release of these elements into the biosphere.

The research priorities and scientific documentation published by the CEC [1] and Angeletti [2] have previously provided a review of the problems related to marine radioecological research. Further I feel that some of the recommendations for future research, formulated in the conclusions of the IAEA Panel on the Effects of Ionizing Radiation on Aquatic Organisms and Ecosystems, remain valid for research on transuranics and, therefore, I bring the following points to your attention.

119

(a)   There is a need for more data on the concentrations and distribution between organs, and even in the same organ, of natural and artifical radionuclides including the transuranics in representative species of natural ecosystems.

A better knowledge of the microdistribution in and between organisms of α-emitting radionuclides is essential if one has to calculate the dose to the different organs in aquatic organisms.

(b)   Owing to the nature of α-emitters, which have a high linear energy transfer (LET) in biological material, their importance in terms of dose rate to organs such as, for example, embryos of fish etc. which may accumulate these radionuclides, can be great. For α-particles with ranges in tissues of the order of a few tens of microns (μm) it is clear that the distribution of the source needs to be known on this scale for an accurate estimation of radiation dose.

(c)   Fresh and brackish waters present fluctuating environmental conditions, in contrast to the more stable marine environment. As coastal regions and inland waters could potentially show the highest biological accumulation of radiocontaminants owing to the relative lack of diluting volume, the related experimental studies in freshwater radioecology must emphasize these fluctuations. Both short-term fluctuations, such as temperature, salinity, organic load, stream velocity, pollution peaks, and long-term trends such as pH-decrease by continuing use of sulphur-containing fuels, could be studied.

The interaction of other species on the organisms studied can also be considered as an environmental influence. The interspecific interactions can best be approached experimentally in artificial (micro-) ecosystems with a limited amount of species and trophic levels and with adequate methods of modelling. The importance of environmental variables for the prediction as well as for the interpretation of radioecological data should be stressed, e.g. for the influence of rainfall [3] and of diurnal changes [4].

These three points still indicate very pratical guidelines for research needs as formulated in Technical Reports Series No. 172 of the IAEA.

2.   ABIOTIC FACTORS AFFECTING BIOACCUMULATION

2.1. Physico-chemistry of the radionuclides

Quite complicated physico-chemistry of e.g. plutonium in the aquatic environment is not an exceptional fact in radioecology; an introduction to these difficulties was the study of ruthenium [5, 6].

The growing awareness of the importance of the physico-chemical forms (ultra-filtrability, filtrability, colloid formation, oxidation state, association with other colloids, polarity, etc.) can also be seen e.g. in the study of the toxicity of metals in the aquatic environment [7]. The aquatic radioecologists expect that a good deal of the hitherto unexplained results and not well understood variability in experiments with transuranic elements will be clarified when more is known about the physico-chemistry of the transuranics. Some results from our laboratory indicate concentration factors of 10^4 for ^{241}Am by adsorption phenomena in a freshwater snail, and differences in bioaccumulation of one order of magnitude in closely related surface waters were observed [8].

These physico-chemical aspects have also to be integrated in modelling [9]. They could play a role in the interaction of bioaccumulation in the presence of heavy metals. They can also intervene during the extremely variable conditions encountered in irrigated systems [10] that overlap the radioecology of aquatic and terrestrial systems. Radioisotopes of elements that have a well-defined physiological role, or are known to closely simulate elements of biological importance, have the advantage for biological radioecologists to stimulate their interest in the mechanisms underlying their bioaccumulation as a fundamental aspect of physiology. Notwithstanding this, even for such isotopes as cobalt, manganese and strontium [9, 11], a very wide variation in bioaccumulation factors is reported in different natural water bodies; physico-chemical speciation may play an important role in this variation. It is important that studies on the bio-accumulation of transuranic elements be undertaken in controlled laboratory conditions, making use not only of so-called 'synthetic' waters in which the solutes are quite well known, but also of 'natural' water for which a precise analysis should be available. Such experiments greatly increase the comparability of results. They also permit a better extrapolation of some laboratory conditions to the real disposal conditions in the biosphere, in both freshwater and marine systems. The study of the physico-chemical state becomes very complex when the contaminated waters undergo the transition from freshwater to the more marine conditions of estuaries. It remains an open question as to whether the generation of nanometer particles by ageing of plutonium solution (see e.g. Ref.[12]) is of significance for the biological availability of transuranics in aquatic ecosystems.

2.2. Contrast between freshwater and marine ecosystems

Radioecologists are used to considering the isotopic dilution in marine environments as being much higher than in freshwater, sometimes leading to lower concentration factors in the marine organisms. However, this does not hold for the transuranics and the different degrees of bioaccumulation of plutonium and americium in these different environments are not clearly understood. The americium to plutonium ratio in marine algae is very variable and it is sometimes considered that plutonium is more available in the marine than in freshwater milieu [13–15]; but, in other instances, the opposite occurs [16, 17]. The fate of

transuranics in marine environment has been studied much more extensively
than in freshwater systems. Only recently has this begun to change [18].
Most of the information on freshwater ecosystems is from relatively eutrophic
lakes such as Lake Ontario [13]. Americium dissappears relatively rapidly in the
marine environment; according to Murray et al. [18] at a distance of 500 km
from Windscale the Pu/Am ratio decreases by 30 times. This probably occurs by
association of the americium with suspended material and may thus gradually
change the relative availability of both transuranics. The ratio of plutonium in
soft parts to shell of some marine molluscs also changes as a function of the
distance from the release point, suggesting a change in bioavailability of this
isotope [19]. It has been indicated that plutonium from nuclear fuel reprocessing plants could be 10 to 100 times more available than fallout plutonium [20],
and for this phenomenon no valid explanation is known.

2.3. Influence of temperature, other pollutants and eutrophication

The rise of temperature of water affects the solubility of oxygen as well as
the rates at which metabolic processes occur in poikilothermic animals. It also
has an influence on the length of life cycles and can thus cause a shift in the
succession of different developmental stages of different species. This latter
factor can have an important influence on the availability of prey for predators
and can change the fundamental links in an ecosystem. It is not always very clear
if the reported differences in bioaccumulation factors for radionuclides at
different temperatures are primarily due to a decrease of oxygen supply in the
aquatic medium or to an increase in metabolic rates of the animals. It is generally
known that per 10°C the uptake of radionuclides will not increase by more than
a factor of 2 or 3, and quite often even less [5, 21]. Studies such as those at the
Savannah River Plant [22] could, when transposed to conditions more likely to be
encountered in Europe, give useful information on this question. The impact of
thermal releases on species normally living in (sub)tropical aquatic ecosystems
which are generally touching their upper temperature limits quite closely, could
be much more severe than for species living in colder waters [23].

The metabolic interaction of metals is also worth while to be studied from
the point of view of interference with the bioaccumulation of transuranics. Such
interactions are quite well known in human medicine and toxicology of mammals,
and a classical example are the zinc/cadmium and the zinc/selenium interactions.
Much less is known about these interactions in freshwater and marine animals, and
nothing is known about such interactions in the bioaccumulation of radionuclides
such as the transuranics. Eutrophication can also directly or indirectly (by
stimulation of algae bloom and thus by increase of the adsorption surface for
transuranics in the water) influence the cycling of transuranics. Studies on meso-
or oligotrophic conditions are extremely scarce [8], compared with those in
eutrophic lakes [15, 17].

3.   BIOTIC FACTORS AFFECTING BIOACCUMULATION

3.1. Microorganisms in sediments

The role of sediments as a trap for most radionuclides has long been known. In contrast with this, I feel that there is a serious lack of knowledge on the possible mobilizing effects of the biota, and of long-term changes in water quality, on this reservoir of nuclides. I could not find experimental studies in Europe on the influence of microbiological actions on the radionuclides trapped in river sediments.

Together with the microbiological action, but probably an order of magnitude less important, are acting benthic organisms on the solubilization and on the resuspension of the sedimented radioactivity. The microbial action is given as an example by Saas [24], when following the migration of cobalt in soil. This migration is stopped when the cobalt is bound by microbes in the soil. After microbial inhibition, for example by antibiotics, the cobalt complex continues to migrate and may become reincorporated into the ground water. Plutonium binding by cellular and exocellular components was also demonstrated in fungal cultures [25]. In soil an organic and mobile plutonium ligand with a molecular weight of about 1 million has been identified [26]. Obviously, radioecologists urgently need the collaboration of microbiologists interested in the mineral cycling by bacteria. One of the recent reviews on bacterially mediated transformation of metallic compounds [27] could give useful inspiration in this respect. Notwithstanding the important amount of radionuclides trapped by sediments, some controversy exists about the importance of these radionuclides as a source for bioaccumulation. Thus, caesium [28, 29] as well as plutonium and americium [30] present in the sediments are sometimes quoted as being of low importance for uptake by sediment dwelling organisms. Suspended particles would have a different influence on different isotopes [31] and, thereby, induce a different availability for biological fixation. The exchange surface between water and solid materials will be very important between flood plains and streams [32]; some European river systems offer exceptionally good areas for studying this phenomenon. Benthic organisms could, by feeding and by reworking the sediments, change the stratification of the radioactive deposit [33] and resuspended americium and plutonium [15], with possibly subsequent physico-chemical changes. Such resuspensions have been recognized as important for the availability of americium and plutonium in Lake Michigan biota [15] and in the Great Lakes [17], in which benthic food chains had higher plutonium levels. The occurrence of plutonium in the lungs of small rodents along the Los Alamos canyon was also attributed to resuspension related to liquid waste discharges in this area [34].

It is common practice in agronomy to relate the bioavailability of metals
with their extractability by a number of solvents, but a lack of such relationship
was found for the bioavailability of ^{109}Cd from sediments [35]. This quite
exceptional finding is also worth while checking and extending to other metals or
nuclides, and to other sediments.

3.2. Accumulation processes

The fact that living organisms can fix, in a more or less stable way, some
radioactivity present in the environment can be due to metabolic pathways, to
surface adsorption, or to ingestion of contaminated food or sediments. The
result is the contamination of the organism; the amount of radioactivity thus
fixed in or on the organism being expressed by the ratio of the activity per unit
wet weight of the organism compared to the radioactivity present in the same
unit of weight of the environmental water or sediment. Obviously this concentration factor is depending on the time of contact of the organism with the
radioactivity source as well as being a function of a number of other variables
such as the physiological condition of the organism, its surface to volume ratio,
its feeding habit, and so on.

Owing to the complex nature of transuranic physico-chemistry, the study
of accumulation mechanisms of these radionuclides is clearly necessary;
consideration of the problems in the extrapolability of laboratory results to
field situations is also of importance.

In this context the potential usefulness of work on indicator organisms has
to be stressed. This should allow comparisons on the uptake of transuranium
nuclides by the same organism in different environmental conditions.
The intercomparability of work of different laboratories will also increase if
some research on indicator species could be undertaken. *Mytilus edulis* has been
reported as being a valuable indicator organism for the uptake of the transuranics.
Water cress (*Rorippa*) and dragonfly (*Libellula*) larvae were reported to have high
accumulation factors for transuranics [36]. For freshwater biota the gastropod
Lymnaea stagnalis L. has a very broad ecological range and is also quite easy to
breed in the laboratory. It also displays relatively high concentration factors
of about 1000 for americium and plutonium.

Although very little is known about the physiological mechanisms by which
bioaccumulation of the transuranics occurs, it appears that food chains do not
show a marked increase in concentration factors. The moulting of shrimps
results in the loss of much of the plutonium fixed by or on these animals. The
accumulation of the quadrivalent or hexavalent plutonium forms was equally
efficient in some marine polychaeta worms. No differences between the different
isotopes were found if the chemical form of the different isotopes was identical [37].

In general, concentration factors are quite similar in freshwater and marine environments; they are about 1000 for phytoplancton, algae and molluscs and they decrease to 10 for fish; these concentration factors become one order of magnitude lower when they are related to the total plutonium concentrations in the water as opposed to the filtratable plutonium concentrations in the water [38].

The very high concentration factors found in the byssus of *Mytilus edulis* is similar to ruthenium. Biochemical analysis of the material constituting this byssus could give some indication of the kind of organic structure exhibiting such a high affinity for plutonium. Similar high concentration factors for americium were observed in the newly formed border on the shell of freshwater gastropods [8].

At present, it is not clear what kind of metabolic pathway, or what kind of biologically important mineral, if any, is simulated by the transuranics bio-accumulation processes. An important field of research for physiologists appears to present itself here.

Such research could also throw some light on the relative importance of uptake of the transuranics from food and directly from the water.

The observations of Guary [39] are interesting in this respect as they point to some parallelism between plutonium and calcium fixation in marine organisms. Their interpretation of the inverse relationship between concentration factors and trophic levels of marine organisms requires further investigation. It was implied that some of the plutonium swallowed by bottom dwellers such as *Pleuronectes platessa* could become assimilated. More research could throw light on this very important observation, as well as on the high concentration factors obtained in the starfish *Asterina gibbosa.*

It is unfortunate that some of the work done on concentration factors in field studies is hampered by the restricted number of animals. Methods used for sampling to check the contamination by heavy metals in littoral and pelagic marine organisms could be useful in this respect [40] to yield acceptable statistics with a minimum number of animals.

4.   GENERAL CONCLUSIONS

Physico-chemical research on the speciation of the transuranics in aquatic environments, together with a better knowledge of the physiological mechanisms intervening in the assimilation of these nuclides by the aquatic organisms, is needed to get a better insight into their bioaccumulation. Adsorption phenomena have to be explained also on a physical and chemical basis. The relative importance of the radiocontamination by transuranics in the actual situation of environmental radiocontamination clearly has to be properly evaluated.

REFERENCES

[1] COMMISSION OF THE EUROPEAN COMMUNITIES, Radiation Protection Programme
     1980—1984: Research and Scientific documentation, XII/1067/79, CEC, Brussels
     (Oct. 1979).

[2] ANGELETTI, L., Les transuraniens: Propriétés physicochimiques et comportement
     dans l'environnement, Rep. CEA-R-4987 (1979).

[3] SHURE, D.J., GOTTSCHALK, M.R., "Cesium-137 dynamics within a reactor effluent
     stream in South Carolina", CONF-750503-27 (1975) 27 pp.

[4] YOUSEF, Y.A., PADDEN, T.J., GLOYNA, E.F., Diurnal changes in radionuclides uptake
     by phyto-plankton in small scale ecosystems, Water Res. 9 2 (1975) 181.

[5] VANDERBORGHT, O., VAN PUYMBROECK, S., Initial uptake, distributions and loss
     of soluble ^{106}Ru in marine and freshwater organisms in laboratory conditions, Health
     Phys. 19 (1970) 801.

[6] BERG, G.G., GINSBERG, E., Partition of Ru-106 between the fresh water environment
     and crayfish, Health Phys. 30 (1976) 329.

[7] DRISCOLL, C.T., BAKER, J.P., BISOGNI, J.J., SCHOFIELD, C.L., "Aluminium
     speciation and its effect on fish in dilute acidified waters", Ecological Impact of Acid
     Precipitation, Int. Conf. Sandefjord, Norway, 11—14 Mar. 1980.

[8] THIELS, G.M., VANGENECHTEN, J.H.D., VANDERBORGHT, O., "Biological
     availability of the transuranic element ^{241}Am in oligotrophic and mesotrophic surface
     waters: combined effect of acidity and phosphate", Biological Implications of
     Radionuclides Released from Nuclear Industries (Proc. Symp. Vienna, 1979), IAEA,
     Vienna (1979) 363.

[9] VANDERPLOEG, H.A., PERZYCK, D.C., WILCOX, W.H., KERCHER, J.R., Bio-
     accumulation Factors for Radionuclides in Freshwater, Rep. ORNL 5002 (1975).

[10] MYTTENAERE, C., MOUSNY, J.M., DABIN, P., Transfer of radioactive and chemical
     pollutants into irrigated rice fields, Radioprotection 10 4 (1975) 235.

[11] DUGUID, J.O., Annual Progress Report of Burial Ground Studies at Oak Ridge National
     Laboratory, Rep. ORNL 5141 (Oct. 1976) 62 pp.

[12] STRADLING, G.N., LOVELESS, B.W., HAM, G.J., SMITH, H., The biological
     solubility in the rat of Pu present in mixed Pu-Na aerosols, Health Phys. 35 (1978) 229.

[13] LIVINGSTON, H.D., BOWEN, V.T., Contrasts between the Marine and Freshwater
     Biological Interactions of Plutonium and Americium, COO-3563-33 CONF-7510121-1
     (1975).

[14] LIVINGSTON, H.D., BOWEN, V.T., Americium in the Marine Environment: Relationships
     to Plutonium, CONF-750672-2 (1976).

[15] WAHLGREN, M.A., ALBERTS, J.J., NELSON, D.M., ORLANDINI, K.A., "Study
     of the behaviour of transuranics and possible chemical homologues in Lake Michigan
     water and biota", Transuranium Nuclides in the Environment (Proc. Symp. San Francisco,
     1975), IAEA, Vienna (1976) 9.

[16] PENTREATH, R.J., LOVETT, M.B., Occurrence of plutonium and americium in plaice
     from the north eastern Irish Sea, Nature (London) 262 (1976) 814.

[17] EDGINGTON, D.N., WAHLGREN, M.A., MARSHALL, J.S., "Behaviour of plutonium in
     aquatic ecosystems: Summary of studies in the Great Lakes", 8th Int. Conf. Environ-
     mental Toxicology (MILLER, M.W., STANNARD, J.N. Eds), Ann Arbor Science
     Publishers Inc. (1976) 45.

[18] MURRAY, C.N., KAUTSKY, H., HOPPENHEIT, M., DOMIAN, N., Actinide activities
     in water entering the northern North Sea, Nature 276 (1978) 225.

[19] GUARY, J.C., FRAIZIER, A., Etude comparée des teneurs en plutonium chez divers mollusques de quelques sites littoraux français, Mar. Biol. **41** (1977) 263.

[20] PILLAI, K.C., MATHEW, E., "Plutonium in the aquatic environment: Its behaviour, distribution and significance", Transuranium Nuclides in the Environment (Proc. Symp. San Francisco, 1975), IAEA, Vienna (1976) 25.

[21] VANDERBORGHT, O., VAN PUYMBROECK, S., Kinetics of the direct uptake of Ca and Sr ions from water by freshwater gastropods: influence of temperature, Environ. Physiol. **1** (1971) 83.

[22] SAVANNAH RIVER ECOLOGY LABORATORY, Annual Report of Ecological Research SREL-6 (May 1976) 95 pp.

[23] PATEL, B., BALANI, M.C. et al., "Impact of thermal and radioactive effluents on a tropical nearshore ecosystem", Combined Effects of Radioactive, Chemical and Thermal Releases to the Environment (Proc. Symp. Stockholm, 1975), IAEA, Vienna (1975) 17.

[24] SAAS, A., GRAUBY, A., "Techniques de déterminations rapides des effets de synergie radionuclides-pollutants", Combined Effects of Radioactive Chemical and Thermal Releases to the Environment (Proc. Symp. Stockholm, 1975), IAEA, Vienna (1975) 145.

[25] ROBINSON, A.V., GARLAND, T.R., SCHNEIDERMAN, G.S. et al., "Microbial transformation of a soluble organoplutonium complex", Biological Implications of Metals in the Environment — ERDA Symp. Ser. No. 42 (1975) 52.

[26] GARLAND, T.R., WILDUNG, R.E., "Physicochemical characterisation of mobile plutonium species in soils", Biological Implication of Metals in the Environment — ERDA Symp. Ser. No. 42 (1975) 254.

[27] EHRLICH, H.L., Inorganic energy sources for chemo-lithotrophic and mixotrophic bacteria, Geomicrobiol. J. **1** (1978) 65.

[28] VANDERPLOEG, H.A., BOOTH, R.S., CLARK, F.H., "Specific activity and concentration model applied to cesium-137 movement in a eutrophic lake", 4th Natl. Symp. Radioecology, CONF-750503-7 (1975) 31.

[29] EMERY, R.M., KLOPFER, D.C., GARLAND, T.R., WEIMER, W.C., Ecological Behaviour of Plutonium and Americium in Freshwater, Rep. BNWL-SA-5346 (Mar. 1975) 38 pp.

[30] BEASLEY, T.M., FOWLER, S.W., Plutonium and americium: uptake from contaminated sediments by the polychaete worm *Nereis diversicolor*, Mar. Biol. **38** (1976) 95.

[31] HARRISON, F.L., WONG, K.M., HEFT, R.E., Role of Water and Particulates in Radionuclide Accumulation in the Oyster *Crassostrea gigas*, Rep. SCRL-76570 (Res. 1) (1975) 33 pp.

[32] PLATT, R.B., RAGGDALE, H.L., SHURE, D.J., Ecological Behaviour and Effects of Radionuclides in Southeastern Ecosystems, Progress Report ORO-2412-65 (1976) 154 pp.

[33] HORIKOSHI, MASUOKI, Feeding Habit and Mode of Living Benthic Organisms, in Relation to the Radioecology, Rep. NIRS-M-10 (Oct. 1975) 36. (In Japanese.)

[34] HAKONSON, T.E., BOSTICK, N.V., ^{137}Cs and plutonium in liquid waste discharge areas at Los Alamos", Radioecology and Energy Resources (CUSHING, C.E., Ed.), Halsted Press (1976) 40.

[35] LUOMA, S.M., JENNE, E.A., "Factors affecting the availability of sediment-bound cadmium to the estuarine, deposit-feeding clam *Macoma balthica*", Radioecology and Energy Resources (CUSHING, C.E., Ed.), Halsted Press (1976) 283.

[36] EMERGY, R.M., KLOPFER, D.C., GARLAND, T.R., WEIMER, W.C., "Ecological behaviour of plutonium and americium in a freshwater pond", Radioecology and Energy Resources (CUSHING, C.E., Ed.), Halsted Press (1976) 74.

[37] FOWLER, S., HEYROUD, M., BEASLEY, T.M., "Experimental studies on plutonium kinetics in marine biota", Impacts of Nuclear Releases into the Aquatic Environment (Proc. Symp. Otaniemi, 1975), IAEA, Vienna (1975) 157.

[38] HETHERINGTON, J.A., JEFFERIES, D.F., LOVETT, M.B., "Some investigations into the behaviour of plutonium in the marine environment", Impacts of Nuclear Releases into the Aquatic Environment (Proc. Symp. Otaniemi, 1975), IAEA, Vienna (1975) 193.

[39] GUARY, J.C., FRAIZIER, A., Influence of trophic level and calcification on the uptake of plutonium observed in situ in marine organisms, Health Phys. 32 (1977) 21.

[40] MARTIN, J.H., Bioaccumulation of Heavy Metals by Littoral and Pelagic Marine Organisms, Rep. EPA-600/3-79-038 (1979).

CHEMICAL SPECIATION AND BIOAVAILABILITY OF TRANSURANICS FOR A FRESHWATER SNAIL (*Lymnaea stagnalis* L.)*

G.M. THIELS**, C.N. MURRAY, J. RADE
Commission of the European Communities,
Joint Research Centre,
Ispra Establishment,
Chemistry Division,
Ispra, Varese,
Italy

Abstract

CHEMICAL SPECIATION AND BIOAVAILABILITY OF TRANSURANICS FOR A
FRESHWATER SNAIL *(Lymnaea stagnalis* L.).

It is now becoming clear that the determination of the physico-chemical forms of transuranic elements is an important step in assessing their behaviour at very low environmental levels. Data from both simulated environmental systems as well as in-situ investigations have shown the necessity of understanding the source term of contamination, which probably plays a major role in the long-term distribution of these elements. In the present paper an experimental procedure is outlined, which allows a more extensive investigation into some aspects of the biogeochemical behaviour of two transuranics: ^{237}Pu and ^{241}Am. Two chemical methods were applied to a study of the freshwater snail *Lymnaea stagnalis* L.. Data were obtained on the uptake and retention patterns of different oxidation states of ^{237}Pu and ^{241}Am at the organ and cellular levels of the pond snail. An attempt was made to relate the environmental chemistry of both radionuclides to the fixation in *L. stagnalis*.

1. INTRODUCTION

To evaluate the capacity of an aquatic ecosystem to accept radionuclides safely, knowledge is required of the concentration factors in organisms and of the parameters which influence these concentration factors. Information concerning the bioavailability of the transuranic elements plutonium and americium for freshwater organisms is scarce and limited to a few species. For large aquatic systems Bowen and Noshkin [1], Bowen [2], Yaguchi et al. [3], Waller et al. [4] and Wahlgren et al. [5] have published data, pertaining mainly to the biota in the Great Lakes, USA. In rare instances transurancis may occur at higher levels in freshwater ecosystems, such as in isolated ponds and lagoons used to receive low-level waste from plutonium processing operations. Emery et al. [6−8] and Johnson et al.

---

* This work was partly performed with a specialization grant of the CEC.
** CEC Bursar, on leave from the University of Antwerp, Belgium.

[9–11] have reported on the ecological distribution of plutonium and americium in these aquatic systems. Finally, some laboratory data for simulated freshwater systems are available as described by Giesy and Paine [12], Murray et al. [13] and Thiels et al. [14, 15]. The factors governing the physico-chemical behaviour of these transuranics in aqueous systems are poorly understood [16–18].

The possibility of increased input of these man-made elements into the freshwater environment is apparent with the increased production of radiotoxic waste by the nuclear industry. Investigation of the biogeochemical aspects of these radionuclides at low environmental levels is thus of fundamental importance in assessing the potential hazards to man through the food chain.

In the present paper an experimental methodology is outlined, which allows a more extensive investigation into some aspects of the biogeochemical behaviour of two transuranic elements. A chemical separation technique, yielding information on particle and complex forms of plutonium and americium in aqueous solutions, was applied to a study of the freshwater snail, *Lymnaea stagnalis* L.. Data were obtained on the uptake and retention patterns of different oxidation states of ^{237}Pu and ^{241}Am at the organ and the cellular levels in the pond snail. An attempt has been made to relate the environmental chemistry of both radionuclides to the fixation in *L. stagnalis* and some preliminary results will be briefly discussed.

2.   ANALYTICAL PROCEDURES

A combination of methods is available to determine different forms of plutonium (III), (IV), (VI) and americium (III). In the present study a chemical separation technique together with thenoyltrifluoroacetone (TTA) solvent extraction was used to characterize the various complex forms and valencies in artificially contaminated samples of Lake Maggiore water.

2.1. Production of the chemical forms of ^{237}Pu and ^{241}Am

In order to produce initially known oxidation states of plutonium and americium, use has been made of the method described by Murray and Fukai [19]:

Plutonium (III): an aliquot of stock ^{237}Pu solution (≈ 0.8 μCi) was evaporated to dryness with 2 ml of hydroxylamine hydrochloride solution (50 mg $NH_2OH \cdot HCl$/ 2 ml H_2O) and the residue dissolved in 1.5N HCl;

Plutonium (IV): an aliquot of stock ^{237}Pu solution (≈ 0.8 μCi) was evaporated to dryness with 2 ml 16N HNO_3 in which solid $NaNO_2$ was dissolved. The residue was then dissolved in 1.5N HCl;

Plutonium (VI): an aliquot of stock ^{237}Pu solution (≈ 0.8 μCi) was evaporated to dryness with 2 ml concentrated perchloric acid ($HClO_4$) and the residue dissolved in 1.5N HCl;

Americium (III): a solution of ^{241}Am ($\approx$2.2 μCi) was evaporated to dryness and the residue dissolved in 1.5N HCl.

2.2. Chemical separation technique

In order to determine particle and complex forms of plutonium and americium, an analysis train, developed by the JRC (Ispra), was used. This analysis train consists of the following components:

0.45 μm filter (Sartorius Membranfilter SM 11306) followed by a 0.01 μm filter (Sartorius Membranfilter SM 11318) to determine the formation of insoluble species;

cation resin (BioRad 50 W X 8, 100–200 mesh);

anion resin (BioRad AG 1 X 8, 100–200 mesh);

eluate collector to determine the presence of non-ionic soluble species.

2.3. TTA-solvent extraction

The method developed by the JRC [17] for the determination of the oxidation states of plutonium and americium in aqueous media is based on the work of Foti and Freiling [20], using the organic solvent 2-thenoyltrifluoroacetone (TTA) dissolved in xylene. In this method it has been shown that Pu (III) is extracted at pH 4.3 and not at pH 0.3, Pu (IV) is extracted at both pH values and Pu (VI) is not extracted at either. Am (III) behaves as Pu (III), being extracted at pH 4.3 but not at pH 0.3.

2.4. Measurement of ^{237}Pu and ^{241}Am by gamma spectrometry

The radioactivity due to ^{237}Pu and ^{241}Am was determined with a double NaI(Tl) detector (Packard Autogamma Scintillation Spectrometer, A 5320); both isotopes are efficiently measured with this system (80% and 29% detector efficiency for plutonium and americium, respectively). The details of the gamma spectrum of ^{237}Pu have been discussed by Murray and Fukai [19].

2.5. Biochemical method

In order to obtain some information on the distribution of actinides in the cellular components of *L. stagnalis,* a short test was carried out on the hepato-pancreas and the blood of a number of specimens.

After 11 days of uptake the hepatopancreas of six snails were pooled and homogenized in Tris HCl at pH 8. The repartition of the transuranics between the cellular organelles and the soluble cytoplasmatic fractions was determined after their separation by ultracentrifugation at 100 000 g for 90 minutes. To identify the association of plutonium and americium with cellular components

TABLE I. PHYSICO-CHEMICAL COMPOSITION (mg/l) OF LAKE MAGGIORE
WATER ON 11 DECEMBER 1979

pH-value	Conductivity (μS)	Na^+	K^+	Ca^{2+}	Mg^{2+}	Total iron
8.53	186	2.4	1.4	16	3.4	<0.01
Cl^-	NO_3^-	HCO_3^-	CO_3^{2-}	SO_4^{2-}	PO_4^{3-}	Total carbon
2.0	3.0	47	0.0	25	<0.5	13.7

present in the soluble cytoplasmatic fraction, gel-filtration on Sephadex G 25 and
Sephadex G 75 resins in columns of 1 X 10 cm and 2.5 X 100 cm, respectively,
was performed and the radioactivity of the fractions obtained was determined with
a double NaI(Tl) detector.

3.   EXPERIMENTAL SETUP

    Water of Lake Maggiore with a physico-chemical composition shown in
Table I was passed through a 0.45 μm filter. Plutonium (III), (IV) and (VI)
($\approx$0.8 μCi) and Am (III) ($\approx$2.2 μm) were added to separate 1.1 l samples of
filtered water, which were then divided into 11 polyethylene beakers of 100 ml
volume.
    Laboratory-bred snails, 2 to 3 cm long, were placed in water without being
fed at laboratory temperature (21 + 1°C) for one week. After this they were put
in the prepared solutions without food, one single animal per beaker. One beaker
per valency series was kept as control.
    On days 0, 1, 2, 3, 4, 7, 9 and 11 the pH was measured and a sample of the
water (1 ml) was removed to follow the decrease of the radioactivity in solution.
On the same days chemical separations (analysis train) on 5 ml of pooled water
samples of the snails (0.5 ml from each beaker) from each series and of the
controls were performed. Thenoyltrifluoroacetone solvent extractions were
simultaneously undertaken on 2 ml of the pooled samples and controls.
    The radioactivity of the whole snails was measured after 4, 9 and 11 days,
after which six specimens of each valency series were dissected. The percentage
organ distribution of plutonium and americium in terms of activity was calculated.

At the same time the cellular fixation in the hepatopancreas for Pu (III), (IV), (VI) and Am (III) and in the blood for Am (III) was determined.

The remaining specimens of each valency series (1—4 animals) were then used to study their retention capacity for either ^{237}Pu or ^{241}Am. Each snail was placed individually without food in 100 ml uncontaminated filtered lake water on day 11. On days 14, 16, 18 and 22 the water was analysed as described above and the activity of the whole animals again measured. On day 22 the snails were dissected and the percentage distribution for each organ was determined as before.

4.   RESULTS AND DISCUSSION

In order to determine the evolution of Pu (III), (IV), (VI) and of Am (III) in the controls and the solutions containing L. stagnalis, measurements of E_h, pH, oxidation states and chemical species were made. As an example of the results Fig.1 gives the data for Pu (VI) on uptake. Solvent extraction results from the experimental solutions at pH 4.3 show a tendency for Pu (VI) to reduce to Pu (III) (Fig. 1 G); this relation can be described by:

$$y = 48.3 + 4.5x$$

in which the regression coefficient is significantly different from 0 at the probability level of $P < 0.02$. The results of the control (Fig. 1 C) demonstrate a large spread (regression coefficient not significantly different from 0). The chemical species distribution in the solutions containing the snails (Fig. 1 H) displays a large variability. However, it seems that real differences occur between the control and the experimental solutions (Figs 1 D and 1 H): the production of a cationic species in the control is higher at the end of the uptake experiment than in the solutions containing L. stagnalis; the opposite can be observed for the anionic species. The presence of particles and of neutral species seems to be similar for the control and experimental solutions, although the former differs largely at the beginning of the experiment.

The results obtained by TTA solvent extraction and the chemical separation technique (Table II) for the experimental solutions of each plutonium valency series on day 11 show a similarity, possibly indicating the occurrence of a single oxidation form at this time. The whole-body concentration factors of the snails for the three initial plutonium valencies (279 ± 44 SE, 251 ± 23 SE and 264 ± 23 SE, respectively) demonstrate no significant differences at the probability level of $P < 0.05$ after 11 days; this fact may also support the assumption that only one oxidation state (possibly Pu (III)) may have been present.

On day 11 the organ distribution of plutonium and americium in six snails of each valency series was determined (Table III). After transformation of $x_i/n_i = P_i$ into arcsin $\sqrt{P_i}$ for stabilization of variance and normalization,

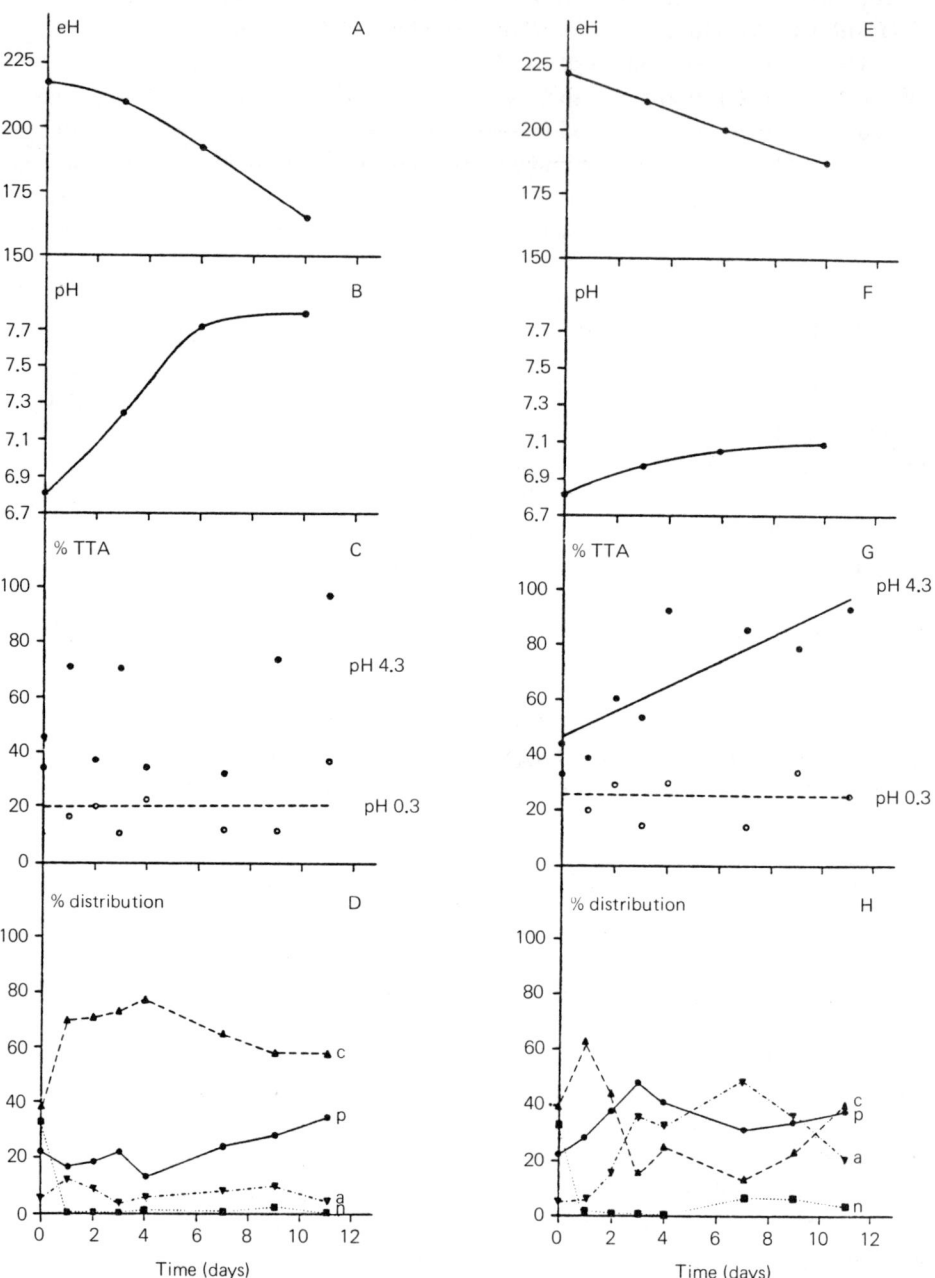

FIG.1.   The E_h and pH values, % TTA and % chemical species distribution for control
(A, B, C, D) and experimental solutions (E, F, G, H) of the Pu (VI) series.
(Symbols: c = cationic fraction; p = particle fraction; a = anionic fraction; n = neutral fraction.)

TABLE II. RESULTS FROM THE TTA SOLVENT EXTRACTION AND THE CHEMICAL SEPARATION TECHNIQUE APPLIED TO THE SOLUTIONS OF THE CONTROLS AND OF THE ANIMALS FOR THE INITIAL THREE VALENCY SERIES ON DAY 11

Method	Control solution			Experimental solution		
	Pu (III)	Pu (IV)	Pu (VI)	Pu (III)	Pu (IV)	Pu (VI)
TTA (%):						
pH = 4.3	95.4	98.7	97.0	87.3	91.2	93.7
pH = 0.3	32.8	64.9	37.3	21.6	31.8	25.5
Species distribution (%):						
particle	63.0	84.6	35.6	37.8	43.6	36.2
cationic	33.3	9.5	57.8	26.9	24.6	40.1
anionic	3.4	4.2	5.4	31.4	28.1	21.1
neutral	0.3	1.7	1.2	3.9	3.7	2.6

TABLE III. PERCENTAGE ORGAN DISTRIBUTION (MEAN ±95% CONFIDENCE LIMITS) OF PLUTONIUM AND AMERICIUM IN TERMS OF ACTIVITY ON DAY 11

Organ	Organ distribution			
	Pu (III) (%)	Pu (IV) (%)	Pu (VI) (%)	Am (III) (%)
Shell	73 ± 8	76 ± 9	80 ± 2	77 ± 2
Shell margin	7 ± 3	7 ± 3	13 ± 3	9 ± 2
G.I.-tract	11 ± 11	8 ± 7	2 ± 2	4 ± 2
Soft tissues	6 ± 5	5 ± 2	3 ± 3	4 ± 2
Hepatopancreas	3 ± 3	2 ± 1	1.5 ± 0.9	4 ± 1
Foot muscle	0.15 ± 0.09	0.3 ± 0.1	0.3 ± 0.2	0.7 ± 0.2
Blood	0.2 ± 0.3	1 ± 1	0.1 ± 0.2	0.9 ± 0.4

Bartlett's test of homogeneity of variance [21], one way analysis of variance and a test for simple and general contrasts were applied to the data. This statistical analysis revealed the following trends:

Shell: no significant differences;
Shell margin: Pu (III, IV) significantly different from Pu (VI) at the probability level of $P < 0.01$;
G.I.-tract: Pu (III, IV) significantly different from Pu (VI) at $P < 0.01$;
Soft tissues: no significant differences;
Hepatopancreas: Pu (IV, VI) significantly different from Am (III) at $P < 0.001$;
Foot muscle: Pu (III, IV, VI) significantly different from Am (III) at $P < 0.001$ and Pu (III) from Pu (VI) at $P < 0.05$.

Since Bartlett's test of homogeneity of variance gave rise to inhomogeneity for the data of the blood, comparison by H-test of Kruskal and Wallis [21] and multiple comparisons according to Nemenyi [21] were applied. This showed that there exists a significant difference between Pu (III) and Am (III) and also between Pu (VI) and Am (III) at the probability level of $P < 0.05$. Thus, when significant differences occur, the higher oxidation state of plutonium distinguishes itself from the lower oxidation states. Uptake is very rapid during the first two days. The differences shown above in the organ distributions related to the various oxidation states thus probably are due to the initial forms of the transuranics added to the water and not to any slow changes in oxidation states or physico-chemical forms, which subsequently occur.

The pattern of uptake per gramme weight for the two radionuclides was found to be: shell margin > shell > whole animal > hepatopancreas > remaining soft tissues > foot muscle > blood. This is in agreement with the data for [241]Am reported by Thiels et al. [14, 15]. The fact that the shell structure accounts for 80% or more of the total body activity (Table III) probably indicates some form of surface adsorption of the actinides. Unpublished autoradiographic results have shown that for a period of up to one week [241]Am is not metabolically incorporated in the shell.

The biochemical tests performed on the hepatopancreas of *L. stagnalis* show that the three oxidation states of plutonium are mainly bound on the cellular membranes ($\approx 70\%$), while $\approx 30\%$ of the activity is found in the soluble fraction. The gel-filtration of the Pu (VI) series demonstrates that in the soluble fraction plutonium is associated with high molecular weight components of undefined nature. This could not be studied for the Pu (III) and (IV) series, since the radio-activity was too low to obtain clear results from the elution profiles. Concerning the fixation of [241]Am in the hepatopancreas, 90% of the radioactivity is bound on the cellular membranes, while only 10% is present in a soluble form. Of the latter, 70% is eluted in fractions corresponding to high molecular weights and only a small percentage is present in a diffusible form. Similar results were obtained from the gel-filtration of the blood, in which only 30% of the americium is dialysable. Thus, in contrast with the fixation at the organ level, a distinct difference between the cellular distribution of ^{237}Pu and ^{241}Am in the hepato-pancreas of *L. stagnalis* has been shown to occur. These preliminary results indicate that further studies using higher activities of both isotopes are necessary to acquire a better understanding of their repartition between cellular organelles and the soluble cytoplasmatic fractions.

The retention capacity of *L. stagnalis* for plutonium and americium was also tried to be determined. Analysis of the oxidation states for each of the plutonium series seems to indicate the production of a (III) valence state. It is not possible to estimate whether this was due to direct loss of this form by the animals or to reduction in situ. The time distribution of the chemical species demonstrated a large variability and did not give rise to clear results.

The percentage organ distribution for both plutonium and americium at day 22 differed from that obtained in uptake (day 11). Under the experimental conditions used some translocation of radioactivity from the soft organs to the shell seems to have occurred. The loss rate of plutonium and americium from the snail into the water appears to be constant with time and the retention for the whole body after 11 days of loss remains about 80% for all series.

5.   CONCLUSION

An attempt has been made to relate a number of environmental, chemical, biological and biochemical parameters which are considered to be of importance in understanding the behaviour of actinides in aquatic ecosystems. It is clear from the complex interactions which have been shown to occur in the simple system described in the present paper that the major bioavailable fractions are still unknown. The preliminary results indicate some of the problems encountered in describing the relationship between the aquatic system and the biological accumulation processes.

ACKNOWLEDGEMENTS

The authors wish to express their gratitude to Dr. M. Hoppenheit for the help in the statistical analysis of the data.

REFERENCES

[1] BOWEN, V.T., NOSHKIN, V.E., "Plutonium concentration along fresh water food chains of the Great Lakes, USA", General Summary of Progress 1972–1973, Woods Hole Oceanographic Institution, Report COO-3568-3 (1973).

[2] BOWEN, V.T., "Plutonium and americium concentrations along fresh water food chains of the Great Lakes, USA", General Summary of Progress 1973–1974, Woods Hole Oceanographic Institution, Report COO-3568-4 (1974).

[3] YAGUCHI, E.M., NELSON, D.M., MARSHALL, J.S., "Plutonium in Lake Michigan plancton and benthos", Radiological and Environmental Research Report ANL-8060, Pt. III, Argonne National Laboratory, Illinois (1973).

[4] WALLER, B.J., NELSON, D.M., MARSHALL, J.S., "Plutonium in Lake Michigan fish", Radiological and Environmental Research Report ANL-8060, Pt. III, Argonne National Laboratory, Illinois (1973).

[5] WAHLGREN, M.A., MARSHALL, J.S., "The behaviour of plutonium and other long-lived radionuclides in Lake Michigan. I. Biological transport, seasonal cycling and residence time in the water column", Impacts of Nuclear Releases into the Aquatic Environment (Proc. Symp. Otaniemi, 1975), IAEA, Vienna (1975) 227.

[6] EMERY, R.M., GARLAND, T.R., The Ecological Behaviour of Plutonium and Americium in a Freshwater Ecosystem. Phase I. Implications of Differences in Transuranic Isotopic Ratios, Battelle Pacific Northwest Laboratories, Richland, Washington, Report BNWL-1879 (1974).

[7] EMERY, R.M., KLOPFER, D.C., "The distribution of transuranic elements in a freshwater pond ecosystem", Eighth Int. Conf. Environmental Toxicology, Radioisotopes in the Aquatic Environment, June 1975, Report BNWL-SA-5424 (1975).

[8] EMERY, R.M., KLOPFER, D.C., McSHANE, M.C., The ecological export of plutonium from a reprocessing waste pond, Health Phys. 34 (1978) 255.

[9] JOHNSON, J.E., WATTERS, R.L., PAINE, D., "The study of plutonium in aquatic systems of the Rocky Flats environs", First Technical Progress Report, Dept. Radiology and Radiation Biology and Dept. Animal Sciences, Colorado Science University, Ft. Collins, Colorado (1972).

[10] JOHNSON, J.E., SVALBERG, S., PAINE, D., "The study of plutonium in aquatic systems of the Rocky Flats environs", Second Technical Progress Report, Dept. Radiology and Radiation Biology and Dept. Animal Sciences, Colorado Science University, Ft. Collins, Colorado (1972).

[11] JOHNSON, J.E., SVALBERG, S., PAINE, D., "The study of plutonium in aquatic systems of the Rocky Flats environs", Final Technical Report, Dept. Radiology and Radiation Biology and Dept. Animal Sciences, Colorado Science University, Ft. Collins, Colorado (1974).

[12] GIESY, J.P., PAINE, D., "Uptake of americium-241 by algae and bacteria", Eighth International Conference on Water Pollution Research, Sydney, Australia, 1976, Prog. Water Technol. 9 4 (1977) 845.

[13] MURRAY, C.N., AVOGADRO, A., LAZZARI, G., "The distribution of actinides in a freshwater microcosm; comparison of simulated input sources", Second Radioecology Symp., Cadarache, June 1979, CEA/CEN Cadarache (1980).

[14] THIELS, G.M., VANGENECHTEN, J.H.D., VANDERBORGHT, O.L.J., "Biological availability of the transuranic element 241-americium in oligotrophic and mesotrophic surface waters: combined effects of acidity and phosphate", Biological Implications of Radionuclides Released from Nuclear Industries (Proc. Symp. Vienna, 1979) 2, IAEA, Vienna (1979) 363.

[15] THIELS, G.M., VANDERBORGHT, O.L.J., Distribution of americium-241 in the freshwater snail *Lymnaea stagnalis* L. and water acidity, Health Phys. (1980) (accepted).

[16] SCHELL, W.R., WATTERS, R.L., Plutonium in aqueous systems, Health Phys. **29** (1975) 589.

[17] JOINT RESEARCH CENTRE, Development of a Method for the Separation and Characterization of Plutonium and Americium Chemical Species in Aquatic Systems, JRC Internal Report (1979).

[18] SALTELLI, A., AVOGADRO, A., BERTOZZI, G., "Assessment of plutonium chemical forms in groundwater", NEA/CEC Workshop Migration of Long-Lived Radionuclides in the Geosphere (1979) (in press).

[19] MURRAY, C.N., FUKAI, R., "Adsorption-desorption characteristics of plutonium and americium with sediment particles in the estuarine environment. Studies using plutonium-237 and americium-241", Impacts of Nuclear Releases into the Aquatic Environment (Proc. Symp. Otaniemi, 1975), IAEA, Vienna (1975) 179.

[20] FOTI, S.C., FREILING, E.C., The determination of the oxidation states of tracer uranium, neptunium and plutonium in aqueous media, Talanta **11** (1964) 385.

[21] SACHS, L., Angewandte Statistik, Springer Verlag, Berlin-Heidelberg-New York (1974).

THE USE OF ISOTOPIC RATIOS IN DETERMINING THE RELATIVE BIOLOGICAL AVAILABILITIES OF TRANSURANIUM ELEMENTS

R.J. PENTREATH
Ministry of Agriculture, Fisheries and Food,
Directorate of Fisheries Research,
Fisheries Radiobiological Laboratory,
Lowestoft, Suffolk,
United Kingdom

Abstract

THE USE OF ISOTOPIC RATIOS IN DETERMINING THE RELATIVE BIOLOGICAL AVAILABILITIES OF TRANSURANIUM ELEMENTS.

The determination of isotopic ratios within marine organisms has been used to estimate relative biological availabilities of various isotopes either in relation to the partitioning of the nuclides in sea water or to the quantities which are introduced into the environment. Such data are particularly valuable in determining relative biological availabilities in situations where the absolute quantities of each radionuclide discharged vary markedly over short periods of time. Difficulties are encountered when nuclides occur as the result of in situ grow-in, such as ^{241}Am resulting from the decay of ^{241}Pu, and the data are also of limited value in that the mechanisms responsible for the observed ratios cannot always be determined. Nevertheless, by comparing isotopic ratios in different organs and tissues it is possible to suggest further lines of experimental investigation. Some examples are given based on $^{239/240}$Pu, ^{241}Am, $^{243/244}$Cm and ^{237}Np concentrations observed in samples taken from the Irish Sea.

1. INTRODUCTION

Although some data are available on the accumulation of plutonium and americium by marine organisms, data on curium and neptunium are generally lacking. There are few areas where such radionuclides are detectable in the marine environment, but one such area is the Irish Sea which contains small amounts as a result of the authorized low-level liquid discharges from the BNFL reprocessing plant at Windscale. The quantities discharged vary considerably from month to month, and dispersion away from the discharge area is also variable. Thus, although transuranium nuclides are detectable in marine organisms, it is often difficult to relate them in such terms as concentration factors because the concentrations in the ambient water are so variable; and, in any case, steady-state conditions are unlikely to exist in the immediate vicinity of the discharge area, where these nuclides occur in concentrations which are easily measured. Some useful data can be obtained, however, if suitable comparisons are drawn between different materials.

141

Before discussing some of the data it is necessary to note that, because of
parent-daughter relationships, a nuclide could also be present as the result of
grow-in from another. In the Windscale area the most complicated relationship
is that of ^{241}Am arising from the decay of ^{241}Pu. In recent years the discharges
of ^{241}Am have declined, whereas those of ^{241}Pu have remained relatively constant.
For example, the grow-in from the ^{241}Pu discharges for the period 1976 to 1978
approximately doubled the direct ^{241}Am input, and this needs to be borne in mind
when comparing ^{241}Am concentrations with those of other nuclides. It is also
important to note that the discussion centres on comparing the ratio of nuclides
in one sample with the ratio of the same nuclides in another. In terms of absolute
concentrations (i.e. in Bq·kg^{-1}), those of $^{239/240}$Pu and ^{241}Am greatly exceed those
of $^{243/244}$Cm and ^{237}Np in all samples because greater quantities of these nuclides
are discharged.

2.   METHODS

All analyses have been made by alpha spectrometry using silicon surface
barrier detectors. The methods used for isotopes of plutonium, americium and
curium are those given by Pentreath and Lovett in Ref. [1], and the method for
^{237}Np is outlined by Pentreath and Harvey [2]. The errors given in the tables
are based on the ± 2σ propagated counting errors.

3.   RESULTS AND DISCUSSION

In the first instance, some comparisons have been drawn between two sessile
organisms: the alga *Fucus serratus* and the limpet *Patella vulgata,* both of which
occur at the same intertidal level at St. Bees Head, north of Windscale. The alga
was chosen because it was presumed that its radionuclide content would reflect
the degree of radionuclide adsorption, and the limpet was chosen on the basis
that its radionuclide content should also reflect its ability to absorb radionuclides,
although some external contamination is inevitable. Only limpets living on vertical,
drained surfaces were selected for analysis. The relative concentrations of
$^{239/240}$Pu, ^{241}Am and $^{243/244}$Cm are given in Table I. With regard to ^{241}Am/$^{239/240}$Pu,
it can be seen that the *Fucus serratus* has a higher quotient than the filtrate water,
but a lower quotient than the particulate matter in suspension; in comparison
with the recent discharges, however, it is approximately the same. The *Patella
vulgata* has a much higher quotient in its total soft parts, higher even than the
particulate matter, and the scraped shell is higher still. The latter is possibly a
reflection of ^{241}Am incorporated in previous years when the discharges of this
nuclide were much greater relative to $^{239/240}$Pu.

TABLE I.   RELATIVE CONCENTRATIONS OF PLUTONIUM, AMERICIUM
AND CURIUM IN SAMPLES COLLECTED AT ST. BEES HEAD, CUMBRIA,
MAY 1978
(Errors are based on $\pm 2\sigma$ counting errors)

Sample	$^{239/240}Pu$ (Bq·kg^{-1} wet)	$\dfrac{^{241}Am}{^{239/240}Pu}$	$\dfrac{^{241}Am}{^{243/244}Cm}$	$\dfrac{^{239/240}Pu}{^{243/244}Cm}$
Fucus serratus	55.3 ± 3.0	0.27 ± 0.02	84 ± 47	307 ± 171
Patella vulgata				
Total soft parts	113.0 ± 6.0	0.72 ± 0.05	97 ± 26	136 ± 37
Scraped shell	11.3 ± 0.8	2.40 ± 0.22	159 ± 76	66 ± 32
Shore-line sea water[a]				
Filtrate ($<$ 0.22 µm)	0.0168 ± 0.0008	0.17 ± 0.02	41 ± 24	240 ± 138
Particulate ($>$ 0.22 µm)		0.54 ± 0.03	83 ± 24	154 ± 44
Discharge				
Previous 4 months		0.28	49	178

[a] Concentrations are in Bq·ltr^{-1}.

Because of the very low levels of $^{243/244}Cm$, the counting errors are rather
large, but in comparing these nuclides with ^{241}Am it is clear that *Fucus serratus,*
Patella vulgata, and the particulate fraction are all enhanced in ^{241}Am relative
to the recent discharges. Comparisons of $^{243/244}Cm$ with $^{239/240}Pu$ are less
conclusive, but there is an indication that the *Fucus serratus* is enhanced in
$^{239/240}Pu$ relative to the discharges, whereas the *Patella vulgata* is not.

Comparisons with ^{237}Np can only be drawn in relation to the discharges
because data on water concentrations at that time are not available. However,
it appears that, relative to ^{237}Np, all other nuclides are enhanced, as can be seen
from Table II.

TABLE II.   CONCENTRATIONS OF PLUTONIUM, AMERICIUM AND
CURIUM RELATIVE TO NEPTUNIUM IN SAMPLES COLLECTED AT
ST. BEES HEAD, CUMBRIA, MAY 1978
(Errors are based on ± 2σ counting errors)

Sample	$\dfrac{^{239/240}Pu}{^{237}Np}$	$\dfrac{^{241}Am}{^{237}Np}$	$\dfrac{^{243/244}Cm}{^{237}Np}$
Fucus serratus	2127 ± 347	581 ± 97	6.9 ± 4.0
Patella vulgata			
Total soft parts	406 ± 39	291 ± 25	3.0 ± 0.8
Discharge			
Previous 4 months	82	23	0.5

A more detailed interpretation of such data as those given in Tables I and II
requires additional information on the period of time over which such accumulation has occurred. Data are also required on the rates of intake of these nuclides
relative to their rates of loss. An attempt to record short-term accumulation into
an internal organ was made by analysing the gonads of the sea urchin *(Echinus
esculentus)*, on the assumption that the radionuclide content of the rapidly
developing gonad tissue would only reflect recent intake — the sea urchin
containing no storage organ with which to produce gonadial growth. The results
are given in Tables III and IV. Comparing ^{241}Am with $^{239/240}Pu$, the gonad has
enhanced ^{241}Am relative to the test (the 'shell'), the water, or the discharge; but
the ^{241}Am in the gonad is not markedly enhanced relative to $^{243/244}Cm$ in other
samples, in fact the quotient of the gonad is lower than that of the test. The
$^{239/240}Pu/^{243/244}Cm$ quotient in the gonad is lower than that of any other sample.
Once again, all other nuclides, in both the gonad and the test, are enhanced relative
to the ^{237}Np quotients in the discharge. Thus the overall order of biological
availability to the echinoid gonad appears to be: ^{241}Am greater than or equal to
$^{243/244}Cm$, each of which are greater than $^{239/240}Pu$, which in turn are all greater
than ^{237}Np. Unfortunately it was necessary to pool the gonads from both sexes
to obtain a sufficiently large sample for analysis, and there may well be differences
in the relative accumulation of the nuclides by testes and ovary.

TABLE III. RELATIVE CONCENTRATIONS OF PLUTONIUM, AMERICIUM AND CURIUM IN THE SEA URCHIN *(Echinus esculentus)* COLLECTED AT ST. BEES HEAD, CUMBRIA, MAY 1978
(Errors are based on ± 2σ counting errors)

Sample	$^{239/240}Pu$ (Bq·kg^{-1} wet)	$\dfrac{^{241}Am}{^{239/240}Pu}$	$\dfrac{^{241}Am}{^{243/244}Cm}$	$\dfrac{^{239/240}Pu}{^{243/244}Cm}$
Echinus esculentus				
Aboral test	18.4 ± 1.0	1.02 ± 0.07	312 ± 208	307 ± 205
Gonad	4.50 ± 0.46	2.98 ± 0.32	103 ± 32	35 ± 11
Shore-line sea water[a]				
Filtrate (< 0.22 μm)	0.0168 ± 0.0008	0.17 ± 0.02	41 ± 24	240 ± 138
Particulate (> 0.22 μm)		0.54 ± 0.03	83 ± 24	154 ± 44
Discharge				
Previous 4 months		0.28	49	178

[a] Concentrations are in Bq·ltr^{-1}.

None of these observations indicate the means by which different radionuclide ratios are attained. Animals can potentially accumulate radionuclides either directly from the water or by absorption from the food. A number of lobsters *(Homarus gammarus)* have been analysed individually and the concentrations of transuranium nuclides in different organs and tissues have been compared. As an example of the results obtained, Table V gives the derived quotients for $^{241}Am/^{239/240}Pu$ in two of these lobsters. Comparisons have been drawn between the quotients obtained in some internal organs — those which are eaten by man — with those of the gills, and also with those of the guts and their contents. All internal organs have higher quotients than the gills and, similarly, all except the digestive gland of the second lobster have much higher quotients than the guts and their contents. Thus in both cases there is enhancement of ^{241}Am relative

TABLE IV.   CONCENTRATIONS OF PLUTONIUM, AMERICIUM AND
CURIUM RELATIVE TO NEPTUNIUM IN THE SEA URCHIN *(Echinus
esculentus)* COLLECTED AT ST. BEES HEAD, CUMBRIA, MAY 1978
(Errors are based on ± 2σ counting errors)

Sample	$\dfrac{^{239/240}\text{Pu}}{^{237}\text{Np}}$	$\dfrac{^{241}\text{Am}}{^{237}\text{Np}}$	$\dfrac{^{243/244}\text{Cm}}{^{237}\text{Np}}$
Echinus esculentus			
Aboral test	354 ± 151	360 ± 153	1.2 ± 0.9
Gonad	1500 ± 523	4467 ± 1495	43.3 ± 19.7
Discharge			
Previous 4 months	82	23	0.5

TABLE V. ^{241}Am/$^{239/240}$Pu QUOTIENTS IN TWO LOBSTERS *(Homarus
gammarus)* COLLECTED NEAR WINDSCALE, OCTOBER 1979
(Errors are based on ± 2σ counting errors)

Sample	$\dfrac{^{241}\text{Am}}{^{239/240}\text{Pu}}$	
	Lobster No. 1	Lobster No. 2
Gill	0.31 ± 0.04	0.46 ± 0.03
Gut + contents	0.77 ± 0.08	1.04 ± 0.07
Digestive gland	19.46 ± 1.66	1.27 ± 0.14
Claw muscle	3.49 ± 0.50	2.57 ± 0.22
Tail muscle	20.26 ± 2.72	33.75 ± 2.88

TABLE VI.   RELATIVE CONCENTRATIONS OF ^{241}Am AND $^{239/240}$Pu IN
TAIL AND CLAW MUSCLE SAMPLES OF TWO LOBSTERS COLLECTED
NEAR WINDSCALE, OCTOBER 1979
(Errors are based on ± 2σ counting errors)

		Lobster No. 1	Lobster No. 2
(A)	$\dfrac{\text{Conc. }^{241}\text{Am in tail muscle}}{\text{Conc. }^{241}\text{Am in claw muscle}}$	2.61 ± 0.17	7.10 ± 0.41
(B)	$\dfrac{\text{Conc. }^{239/240}\text{Pu in tail muscle}}{\text{Conc. }^{239/240}\text{Pu in claw muscle}}$	0.45 ± 0.08	0.54 ± 0.06
	$\dfrac{\text{A}}{\text{B}}$	5.80 ± 1.10	13.15 ± 1.65

either to the gill or to the gut. Of particular interest, however, is the marked
difference between the ^{241}Am/$^{239/240}$Pu quotients of the claw and tail muscle
samples in each lobster, and it is of further interest to see how such differences
have been attained. It is obvious that a change in the derived quotient from one
organ to another can result either from one nuclide being relatively enhanced or
from the other nuclide being relatively depleted. In fact, in this case, it is the
result of both processes as can be seen from Table VI. Thus in each lobster the
concentration of ^{241}Am in the tail muscle is considerably greater than that in the
claw, whereas the concentration of $^{239/240}$Pu in the claw muscle is considerably
greater than that in the tail. As a result, the ^{241}Am/$^{239/240}$Pu quotients in the
tail muscle samples are greater than those in the claw muscle samples by factors
of 6 and 13, respectively.

In order to learn more about such processes it is clearly necessary to have a
time series of observations. This has yet to be done with the lobster, but it has
been done with the plaice *(Pleuronectes platessa)*. In this study [1] the ^{241}Am/
^{238}Pu + $^{239/240}$Pu quotients in different organs of plaice were analysed every three
months over a two-year period and compared with (a) the quotients in the monthly
discharges and (b) the quotients derived from summating the discharge data back
to 1972 (Fig. 1). It was observed that the quotients in the gut contents were
similar to those in the summated ones, and that the muscle tissue samples were
slightly enhanced in ^{241}Am. The liver samples, however, were greatly enhanced
in ^{241}Am, and the quotients reflected those of recent discharges, as did those in
the bone.

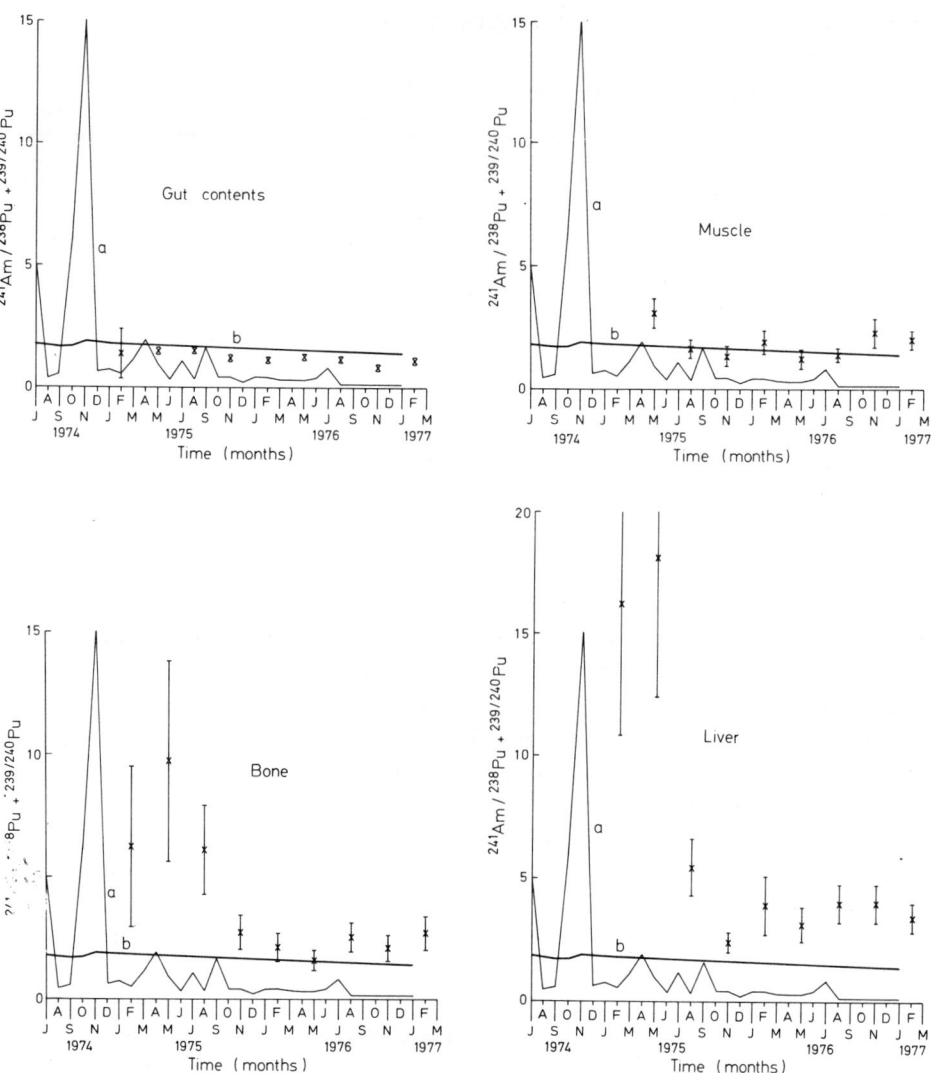

FIG.1.   *Relative concentrations of* ^{241}*Am and* ^{238}*Pu plus* $^{239/240}$*Pu in plaice* (Pleuronectes platessa) *compared with those of the discharge. Line (a): Am/Pu at time of discharge; line (b):* Σ *Am/* Σ *Pu summated back to 1972.*

All these examples are sufficient to demonstrate that, even when all the data normally required are not available, comparative data can be very informative. There are clearly very marked differences in relation to the rates, or extent, of accumulation of the same radionuclide by different organisms; and there are also marked differences in the extent to which different nuclides are accumulated by the same organism. This is hardly surprising, but it is of interest that the opposite also occurs: there are a number of similarities. For example, all of the quotients obtained relative to ^{237}Np imply that this nuclide is accumulated less than the others. Similarly, most of the data imply that ^{241}Am is generally accumulated more than $^{239/240}$Pu — except for *Fucus serratus* — and that this is not merely due to its adsorbent properties but, as demonstrated by the lobster data, is in some cases a result of selective uptake by the animal accompanied by selective deposition into different tissues. Many of the more intriguing questions, of course, particularly those relating to the rates of accumulation, can only be answered by prolonged environmental study and by suitable laboratory and field experiments.

REFERENCES

[1]   PENTREATH, R.J., LOVETT, M.B., Transuranic nuclides in plaice *(Pleuronectes platessa)* from the north-eastern Irish Sea, Mar. Biol. **48** (1978) 19.
[2]   PENTREATH, R.J., HARVEY, B.R., The presence of ^{237}Np in the north-eastern Irish Sea (in preparation).

QUELQUES REMARQUES
SUR LE DEVENIR DES TRANSURANIENS
DANS LE MILIEU AQUEUX

L. ANGELETTI
Association EURATOM/CEA,
Département de protection,
CEA, Centre d'études nucléaires
 de Fontenay-aux-Roses,
Fontenay-aux-Roses,
France

Abstract–Résumé

SOME REMARKS ON THE FATE OF TRANSURANIUM ELEMENTS IN AQUATIC
ENVIRONMENTS.
   This paper compares literature values for transuranium elements from nuclear fallout
and effluent sources in aquatic environments, in fish, sediments and water, for purposes of
prediction of the short- and long-term fate of these radionuclides. Variability of transuranium
values in all environmental samples is noted and explanations are provided for these differences.
It is emphasized, in order to assess adequately the dose to man from nuclear activities, that
better quantitative data are needed on aquatic sediment behaviour and food-chain transfer of
these radionuclides.

QUELQUES REMARQUES SUR LE DEVENIR DES TRANSURANIENS DANS LE
MILIEU AQUEUX.
   Ce mémoire établit une comparaison entre les valeurs données dans la documentation
spécialisée sur les transuraniens provenant de retombées nucléaires et des sources d'effluents
qui se trouvent en milieu aquatique dans les poissons, les sédiments et l'eau pour faciliter les
prévisions du devenir à brève et à longue échéance de ces radionucléides. L'auteur note la
variabilité des valeurs relatives aux transuraniens dans tous les échantillons du milieu et donne
des explications au sujet de ces différences. Afin d'évaluer de façon adéquate les doses à
l'homme résultant des activités nucléaires, l'auteur souligne qu'il est nécessaire de disposer de
données quantitatives plus précises sur le comportement dans les eaux et les sédiments et
le transfert de ces radionucléides par l'intermédiaire de la chaîne alimentaire.

   La connaissance du devenir des transuraniens dans le milieu aqueux a surtout
été illustrée par les données concernant le ^{239}Pu et le ^{240}Pu [1, 2].
   Malgré une abondante littérature concernant sa concentration dans l'eau et
les sédiments ainsi que son transfert dans la chaîne alimentaire, on connaît
actuellement davantage les aspects qualitatifs que quantitatifs du devenir de ces
radioéléments. Cela rend difficile toute évaluation de la distribution de la

TABLEAU I. PLUTONIUM 239 + 240 — TRANSFERT DE L'EAU AUX SEDIMENTS

Référence	Origine du plutonium	Nombre d'échantillons		Sédiments de surface Pu (pCi·kg⁻¹)			Eau de surface Pu soluble[a] (fCi·l⁻¹)			Kd l kg⁻¹ (pCi·kg⁻¹/pCi·l⁻¹)			Observations
		Sédiments	Eau	Min.	Max.	Moyenne	Min.	Max.	Moyenne	Min.	Max.	Moyenne[b]	
[3] Wilson	43 essais d'armes nucléaires à Enewetak Atoll	Nombreux	?	109	3×10^4	964	9	43	26	$2{,}4 \times 10^3$	$3{,}3 \times 10^6$	$3{,}7 \times 10^4$	Pour les sédiments, on donne les valeurs de 15 isoplethes. Le nombre d'échantillons d'eau analysés n'est pas précisé.
[4] Nevissi	23 essais d'armes nucléaires à Bikini Atoll	29	12	400	12×10^4	$3{,}4 \times 10^4$	4	38	27	1×10^4	3×10^7	$1{,}1 \times 10^6$	Echantillons de la station B et C.
[5] Aarkrog	Accident d'arme nucléaire à Thule	37	3	120	6×10^4	1600	1	3	1,6	4×10^4	6×10^7	1×10^6	Le Pu dans le sédiment a été calculé d'après les Kd fournis par l'auteur.
[6] Hetherington	Rejets de l'usine de Windscale dans la mer Iroise	7	7	$1{,}7 \times 10^4$	8×10^4	$2{,}9 \times 10^4$	310	1020	514	$1{,}7 \times 10^4$	$2{,}6 \times 10^5$	$5{,}6 \times 10^4$	Les valeurs concernent les échantillons prélevés entre 1 et 9 km du point de rejet.

[a] Le plutonium soluble est constitué par le plutonium dans les échantillons d'eau filtrée sur des filtres de 0,22 à 0,45 μm.

[b] Les valeurs moyennes du Kd des auteurs [1, 2, 4] sont la moyenne arithmétique. Celle de l'auteur [3] est une moyenne géométrique.

concentration de ces radionucléides dans l'environnement et entraîne de grandes incertitudes quant à l'évaluation des doses aux populations.

Bien que les concentrations actuelles des transuraniens dans le milieu marin soient faibles, tant en valeur absolue qu'en valeur relative, comparées aux autres radionucléides artificiels, il est cependant nécessaire, en raison de leur présence dans les effluents de faible activité rejetés dans les eaux et de leur grande période physique, notamment pour le ^{239}Pu, de bien connaître leur devenir à court terme et de prévoir si possible leur devenir à long terme.

Pour illustrer l'incertitude qui existe quant au devenir du plutonium dans le milieu marin, nous avons réuni dans les tableaux I et II les valeurs de la concentration du plutonium dans les eaux, les sédiments et les poissons, ainsi que les valeurs correspondantes des facteurs de transfert.

Les données du tableau I concernent la distribution, dans les sédiments de surface et les eaux de surface, du plutonium provenant des essais d'armes nucléaires [3—4], d'accidents d'armes nucléaires [5] et des effluents de faible activité [6].

En ce qui concerne le plutonium provenant d'armes nucléaires, on remarque, surtout pour les sédiments, une importante dispersion des valeurs. En fait, lorsque l'on observe pour l'eau une variation maximum d'un facteur 10 environ, on a pour les sédiments une variation d'un facteur 500. Apparemment, la dispersion des valeurs rapportées pour Windscale est très inférieure à la précédente. Mais dans ce cas, on ne sait pas si cette homogénéité dépend davantage d'une situation réelle que du petit nombre d'échantillons analysés.

En ce qui concerne les valeurs moyennes du coefficient de distribution Kd, on ne se sait pas encore si la valeur la plus probable est représentée par 10^4 ou 10^6.

Pour ce qui est de la concentration et des facteurs de transfert du plutonium dans le muscle des poissons pêchés à Enewetak, les valeurs du tableau II font apparaître une dispersion encore bien plus grande que celle que l'on retrouve dans les sédiments. En fait, on observe dans ce cas une variation d'un facteur de 50 000 environ. Les valeurs de la concentration en plutonium dans les muscles des poissons pêchés à Windscale sont plus homogènes et du même ordre de grandeur que celles correspondant à la limite inférieure observée à Enewetak [7].

Lorsque l'on considère la valeur moyenne ou même la valeur médiane des facteurs de transfert observés à Enewetak, on s'aperçoit qu'elles sont du même ordre de grandeur que celles que l'on observe chez les algues. On peut alors se poser la question de savoir si la contamination du plutonium diminue ou non lorsque l'on passe des organismes du premier au troisième niveau trophique. Les données publiées jusqu'à présent sont insuffisantes pour répondre à cette question.

En d'autres termes, on ne dispose pas encore de données suffisantes pour pouvoir comparer le comportement du plutonium d'origines diverses et dans les

TABLEAU II. PLUTONIUM 239 + 240 — FACTEURS DE TRANSFERT DE L'EAU AUX MUSCLES DES POISSONS

Référence	Espèce	Nombre d'échantillons	Concentration dans les muscles (pCi·g⁻¹ poids sec)			Facteur de transfert				Observations
			Etendue	Moyenne	Médiane	Min.	Max.	Moyenne	Médiane	
[3] Wilson	Mulet	25	0,000482–23,1	0,984	0,0145	18,5	$8,8 \times 10^5$	$3,8 \times 10^4$	550	Les facteurs de transfert ont été calculés en prenant la concentration du Pu dans l'eau égale à 26 fCi·1⁻¹.
	Surgeon	28	0,0428 – 0,887	0,0772	0,0280	164	$3,4 \times 10^4$	3×10^3	1000	
	Goatfish	221	0,00161 – 0,0531	0,0130	0,00778	62	2×10^3	5×10^2	300	
	Autres poissons	49	0,000788 – 1,21	0,0700	0,00909	30	$4,6 \times 10^4$	$2,7 \times 10^3$	350	
	Tous les poissons	**123**	**0,000482–23,1**	**0,248**	**0,0126**	**18,5**	$\mathbf{8,8 \times 10^5}$	$\mathbf{9,5 \times 10^3}$	**480**	
[7] Pentreath	Plie	1	–	$2,2 \times 10^{-4}$	–	–	–	0,22	–	Poissons pêchés env. à 5 km du sud du point de rejet. Concentration du Pu soluble: 1000 fCi·1⁻¹. La concentration du Pu est exprimée par rapport au poids frais du poisson.
	Maquereau	1		$2,3 \times 10^{-3}$				2,3		
	Morue	3		$1,2 \times 10^{-3}$				1,4		
	Plie	4	0,00021–0,00131	$9,8 \times 10^{-4}$		0,21	1,31	0,98		
	Tous les poissons	**9**	–	$\mathbf{1,2 \times 10^{-3}}$	–	–	–	1,2	–	

différents milieux aqueux, pour détecter pour un même type de rejet son évolution dans le temps et dans l'espace, pour pouvoir découvrir les types de relations tant entre le milieu et la chaîne alimentaire qu'à l'intérieur de celle-ci.

De ce fait, il est actuellement difficile d'évaluer de façon réaliste les doses aux populations, ce qui constitue, il faut bien le souligner, le but principal des études dans ce domaine.

Il sera donc nécessaire, afin de combler ces lacunes, de connaître, pour un type d'échantillon donné, la distribution de la concentration en plutonium. Cela signifie que la collecte et l'analyse de plus d'une centaine d'échantillons du même type seront nécessaires pour établir les faits. Cela impliquera bien entendu la sélection d'un nombre limité d'échantillons parmi les plus importants pour la chaîne alimentaire, par exemple le plancton et les poissons.

La connaissance de la distribution de la concentration des émetteurs alpha naturels tels que l'uranium et le thorium dans la chaîne alimentaire présente les mêmes lacunes que celles du plutonium [8]. Cela est d'autant plus regrettable que ces radioéléments auraient pu nous donner de précieux renseignements quant au comportement à long terme de différentes formes physico-chimiques du plutonium: l'uranium pour ce qui concerne les formes carbonatées du Pu(VI) et le thorium pour ce qui est des formes hydroxydes et polymères du Pu (IV).

D'une manière générale, et en vue de combler les lacunes existantes, l'étude des émetteurs alpha naturels tels que uranium, thorium, radium, polonium s'avère d'une très grande utilité. De ce fait, elle devrait être si possible entreprise en même temps et pour les même échantillons que ceux choisis pour l'étude des transuraniens car, à partir des résultats ainsi obtenus, il serait possible:
— de mettre en évidence les analogies et les différences et, partant, faire des hypothèses sur le devenir à long terme des transuraniens.
— d'évaluer l'impact des transuraniens par rapport aux émetteurs alpha naturels sur les milieux et aux différentes distances des points de rejet;
— d'évaluer les doses reçues par les populations résultant de la totalité des émetteurs alpha.

Dans cette analyse, et en vue d'évaluer les doses aux populations, on pourrait aussi inclure celle du ^{40}K qui fournit 90% de l'activité bêta due à l'activité des radionucléides naturels et des radioéléments [9].

Des études de laboratoire — tant sur le plan de la chimie que sur celui des transferts — devraient également être menées afin de comprendre l'évolution des formes physico-chimiques des transuraniens dans le milieu aqueux ainsi que leur importance quant au transfert à la chaîne alimentaire.

Dans cette perspective, l'Association EURATOM/CEA a confié au Service des études analytiques de Fontenay-aux-Roses une étude des formes physico-chimiques du plutonium dans l'eau de mer.

Cette étude se propose les buts suivants:
— préparation des formes définies du plutonium: Pu(IV), Pu(VI), polymères;

— contrôle de la fiabilité des méthodes indirectes d'identification des états de
valence du plutonium dans l'eau de mer, c'est-à-dire vérification par la spectrophotométrie d'absorption de la stabilité des différents états d'oxydation du
plutonium au cours des manipulations chimiques de séparation, ainsi que des
aspects quantitatifs des méthodes de séparation chimiques employées;
— étude en fonction du temps de l'évolution des différentes formes physico-
chimiques dans l'eau de mer reconstituée et naturelle.

Les résultats obtenus ainsi que les formes physico-chimiques du plutonium
que l'on aura bien pu caractériser pourraient être mis rapidement à la disposition
de laboratoires de la Communauté européenne pour les études des transferts dans
la chaîne alimentaire.

REFERENCES

[1]   ANGELETTI, L., ANCELLIN, L., BITTEL, R., Aspects pratiques du comportement du
        plutonium dans l'environnement, Radioprotection 12 (1977) 3—26.
[2]   ANGELETTI, L., Les transuraniens — Propriétés physico-chimiques et comportement
        dans l'environnement, Rapport CEA-R-4987 (1979).
[3]   WILSON, D.W., YOOK, C.N.G., ROBINSON, W.L., Evaluation of plutonium at
        Enewetak Atoll, Health Phys. 29 (1975) 599—611.
[4]   NEVISSI, A., SCHELL, W.R., Distribution of plutonium and americium in Bikini Atoll
        lagoon, Health Phys. 28 (1975) 539—547.
[5]   AARKROG, A., Environmental behaviour of plutonium accidentally released at Thule,
        Greenland, Health Phys. 32 (1977) 271—284.
[6]   HETHERINGTON, J.A., JEFFERIES, D.F., LOVETT, M.B., «Some investigations into
        the behaviour of plutonium in the marine environment», Impacts of Nuclear Releases into
        the Aquatic Environment (C.R. Coll. Otaniemi, 1975), AIEA, Vienne (1975) 193—212.
[7]   PENTREATH, R.J., LOVETT, M.B., HARVEY, B.R., IBBETT, R.D., «Alpha-emitting
        nuclides in commercial fish species caught in the vicinity of Windscale, United Kingdom,
        and their radiological significance to man», Biological Implications of Radionuclides
        Released from Nuclear Industries (C.R. Coll. Vienne, 1979) II, AIEA, Vienne (1979)
        227—245.
[8]   CHERRY, R.D., SHANNON, L.V., The alpha radioactivity of marine organisms,
        Rev. En. Atom. 1—2 (1974) 1.
[9]   POLIKARPOV, V.V., Radioecology of Aquatic Organisms, North-Holland Publishing
        Company, Amsterdam (1966).

RADIOLOGICAL STUDY OF A RIVER
RECEIVING RADIOACTIVE LIQUID WASTES
CONTAINING ACTINIDES
First results

M. METAYER-PIRET, K. HOFKENS, J. COLARD
CEN/SCK Mol,
Belgium

R. KIRCHMANN
International Atomic Energy Agency,
Vienna

L. FOULQUIER
CEA, CEN Cadarache,
France

Abstract

RADIOECOLOGICAL STUDY OF A RIVER RECEIVING RADIOACTIVE LIQUID WASTES
CONTAINING ACTINIDES. FIRST RESULTS.
Liquid wastes containing actinides have been discharged for many years according to
legal limits in the Molse Nete, a Belgian river. The fixation rate of plutonium ($\Sigma Pu = ^{238}Pu + ^{239}Pu$
$+ ^{240}Pu$) on suspended matter is about 12% upstream and about 97% downstream from the discharge
point. In 1978, activities sedimented per surface unit 0.1 km downstream ranged from 0.16 to
0.74 for ΣPu, from 0.05 to 0.21 for ^{241}Am and from 0.002 to 0.02 mCi for ^{137}Cs per kg dry
matter per Ci released. Selected data on dredged sediments are also given. ΣPu and ^{241}Am and
^{137}Cs sediment core concentrations are highest 3.6 km downstream (1545, 1759 and 261 nCi
per kg dry matter per Ci released, respectively). ΣPu and ^{241}Am concentrations are similar in
Potamogeton natans L., *P. pectinatus* L. and *Elodea* sp. ($\sim$1400 and 430 nCi per kg dry
matter per Ci released, respectively, 1 km downstream). *Juncus* sp. roots are on an average
6 to 7 times more contaminated than the green parts. *Platyhypnidium* sp. shows high ΣPu
and ^{241}Am concentrations (1200 and 1800 nCi per kg dry matter per Ci released, respectively).
Actinides are also detectable in animals. A spongidae sample presents the highest level measured
(1900 and 2900 nCi per kg dry matter per Ci released for ΣPu and ^{241}Am, respectively). In
agreement with data in the literature, plutonium levels decrease when the trophic level increases.

1. INTRODUCTION

Liquid wastes containing radionuclides such as caesium, antimony,
ruthenium, cobalt, manganese etc. as well as actinides such as plutonium and
americium have been released for many years into the Molse Nete according to
legal limits. In this study the ways of transfer and accumulation in different
aquatic trophical levels and the doses to organisms in these compartments will
be evaluated for the radionuclides released. In addition, an experimental programme

157

will be performed in the laboratory to study the influence of organisms living in the
sediment on plutonium remobilization from the sediment. In this paper the initial
in situ data on actinides for the year 1978 are presented.

2.   MATERIALS AND METHODS

Water and suspended matter: In July 1978, water samples were subject to
continuous centrifugation in a WESTPHALIA centrifuge (type KA 2/06/075 and
KA 2/86/075). Centrifuged water and suspended matter samples were analysed
in order to determine plutonium fixation rates on suspended matter sampled
2.2 km upstream and 0.1 km downstream from the discharge point.

Slime: Traps for slime sampling (0.3 X 0.3 X 0.1 m) placed 2.2 km upstream
and 0.1 km downstream from the discharge point were investigated every three
months.

Dredged sediment: Composite samples (n = 10) of yearly dredged sediment
were collected 0.1, 1 and 3.3 km downstream from the discharge point using a
coring tube type EDELMANBOOR (Eijkelkamp B.V.) (20 cm long and 7 cm in
diameter).

Sediment cores: In June 1978, sediment samples were collected with the
KAHLSICO boat-operated coring tube (No. 217WA070) in the river sedimentation
areas, 2.2 km upstream and 1, 3.6, 17.7 and 35 km downstream from the discharge
point. The composite core samples (n = 3) (3.8 cm in diameter and 20 cm long)
were dried at 105°C and ashed at 600°C.

Plants: In June 1978, plants were collected 2.2 km upstream and 0.1, 1 and
3.6 km downstream from the discharge point, cleaned from attached animals,
dried at 105°C and ashed at 600°C.

Animals: Animals obtained from the collected plants and from superficial
sediments 2.2 km upstream and 0.1, 1 and 3.6 km downstream from the discharge
point were dried at 105°C and ashed at 600°C.

Analysis of actinides [1] :   The ashed (600°C) samples were fused with
potassium carbonate and dissolved in dilute HCl. The alkaline-earth phosphates
were precipitated by addition of H_3PO_4 and concentrated NH_4OH. The
precipitate was dissolved in 8M HCl. The insoluble residue was treated with HF;
the solution was evaporated to dryness, dissolved in 8M HCl and combined with
the HCl solution previously obtained. To avoid any interference, polonium in
8M HCl was adsorbed on an anion exchange resin and the Pu-Am fraction was
eluted with 1M HCl. Plutonium was oxidized to Pu^{4+} with $NaNO_2$ in 7.5M HNO_3,
was adsorbed on an anion exchange resin and eluted with 0.7M HNO_3 containing

---

[1]   The detailed procedure will be published elsewhere by K. HOFKENS, C. HURTGEN.

0.01M hydroquinone. The americium fraction was purified on a cation exchange resin. Americium adsorbed from 2M HNO_3 was eluted with 10M HNO_3. The plutonium and americium fractions were evaporated on stainless steel trays and counted in a 25 cm^3 ZnS alpha counter.

Gamma spectrometry: Gamma spectrometry was performed with a Ge(Li) crystal.

3.   CHARACTERISTICS OF THE MOLSE NETE ECOSYSTEM

The Molse Nete river has an average gradient of 40‰, its average debit is 0.98 $m^3 \cdot s^{-1}$ (0.28 $m^3 \cdot s^{-1}$ in summer and 2.9 $m^3 \cdot s^{-1}$ in winter) [1]. The yearly average of the physico-chemical parameters of water is presented in Table I.

From the elementary analysis of Molse Nete sediments it appears that these sediments are made up of fine siliceous sand (50% of the particles have a diameter between 100 and 200 μm) containing a lot of organic matter (10.5%). The organic matter represents the principal fixation capacity, whereas clays account only for 2%. The total exchange capacity is approximately 18 meq/100 g. The absorbent complex is saturated to 63% of its maximal capacity. Calcium and sulphates are the main water-soluble materials [2].

The aquatic flora is dominated by *Potamogeton* spp., *Elodea* spp., *Callitriche* sp.. Few fishes occur in the area investigated. The benthic crustacea fauna is dominated by *Asellus aquaticus* L.. Molluscs are represented by *Sphaerium* sp., *Physa* sp., *Planorbis* sp., *Lymnaea* sp., *Anisus* sp., *Valvata* sp., *Bithynia* sp. and *Viviparius* sp. The hirudinea group is dominated by *Erpobdella* sp. The tubificidae are represented by *Aulodrilus* sp., *Tubifex* sp. and *Limnodrilus* sp. The most frequent arthropod larvae are chironomidae. The biotic index (Tuffery-Verneaux)[2] is 5.

4.   RESULTS AND DISCUSSION

About 12% of the entire plutonium (n = 1) was fixed to suspended matter upstream and about 97% downstream from the discharge point. These values are similar to those reported in the literature. Singh and Marshall [3] observed that the higher plutonium value in the effluent water released from the Argonne National Laboratory was primarily due to the suspended fraction which contained 99% of the total activity in contrast to 37% of the plutonium activity in the upstream samples. The few fractionated water samples analysed by Bartelt et al. [4]

---

[2]   Carte de la qualité biologique des cours d'eau en Belgique. Ministère de la Santé Publique et de la Famille. Sep. 1979.

TABLE I. YEARLY AVERAGE OF PHYSICO-CHEMICAL PARAMETERS IN THE MOLSE NETE WATER (1978)

Parameter	Unit	Mean ± σ (n = 48)
pH	–	7.20 ± 0.02
Temperature	°C	11.2 ± 0.3
Alcalinity	meq $Ca(HCO_3)_2 \cdot l^{-1}$	6.49 ± 0.08
Sodium	$mg \cdot l^{-1}$	31.2 ± 0.8
Potassium	$mg \cdot l^{-1}$	6.9 ± 0.06
Calcium	$mg \cdot l^{-1}$	49.6 ± 0.7
Magnesium	$mg \cdot l^{-1}$	5.5 ± 0.03
Zinc	$\mu g \cdot l^{-1}$	344 ± 23
Aluminium	$mg \cdot l^{-1}$	0.07 ± 0.009
Iron	$mg \cdot l^{-1}$	0.95 ± 0.04
Chlorides	$mg \cdot l^{-1}$	34 ± 0.3
Nitrates	$mg \cdot l^{-1}$	24.5 ± 2.3
Orthophosphates	$mg \cdot l^{-1}$	0.54 ± 0.07
Sulphates	$mg \cdot l^{-1}$	89.8 ± 2.3
Biological oxygen demand (5d, 20°C)	$mg\ O_2 \cdot l^{-1}$	3.3 ± 0.2
Organic matter	$mg\ C \cdot l^{-1}$	8.7 ± 0.3

indicate that 80% of the plutonium is associated with the suspended sediment fraction from the Miami River watershed and only 20% is found in the fraction filterable through 0.45 μm Millipore filter paper. The yearly ΣPu (sum of Pu(α) isotopes [$^{238}Pu + ^{239}Pu + ^{240}Pu$]), ^{241}Am and ^{137}Cs mean concentrations of slime trapped 0.1 km donwstream from the discharge points are given in Table II, values being highest during the third quarter of 1978. It remains to be seen whether this observation may be explained by the nature and granular size of these slimes.

The activities sedimented per surface unit range from 0.16 to 0.74, from 0.05 to 0.21 and from 0.002 to 0.02 mCi per kg dry matter per Ci released for ΣPu, ^{241}Am and ^{137}Cs, respectively.[3]

ΣPu, ^{241}Am and ^{137}Cs concentrations were measured in dredged sediments at different distances from the discharge point. The values are presented in Table III.

---

[3]   1 Ci = 3.7 × 10^{10} Bq.

TABLE II. ΣPu, ^{241}Am AND ^{137}Cs CONCENTRATIONS IN MOLSE NETE SLIME FROM SLIME TRAPS SITUATED 0.1 km DOWNSTREAM FROM THE DISCHARGE POINT DURING THE FOUR QUARTERS OF 1978
(10^{-6} Ci per kg dry matter per Ci released)

Period	ΣPu [a]	^{241}Am [a,b]	^{137}Cs [a,b]	Pu/Cs	Pu/Am
Jan.–Feb.–Mar.	24.2	4.6	0.15	161.3	5.3
Apr.–May–June	14.6	7.1	0.50	29.2	2.1
July–Aug.–Sep.	66.8	18.8	1.70	39.3	3.6
Oct.–Nov.–Dec.	ND	11.9	0.46	–	–
$\bar{X} \pm 1\sigma$	35.2 ± 22.8	10.6 ± 5.4	0.7 ± 0.6		

Note: ΣPu = ^{238}Pu + ^{239}Pu + ^{240}Pu; ND = Not yet determined; 1 Ci = 3.7 × 10^{10} Bq.

[a] $\dfrac{\text{Downstream value} - \text{Upstream value}}{\text{Release}}$

[b] Measured by gamma spectrometry (GeLi).

TABLE III. ΣPu, ^{241}Am AND ^{137}Cs CONCENTRATIONS IN DREDGED SEDIMENTS OF MOLSE NETE RIVER AS A FUNCTION OF THE DISTANCE FROM THE DISCHARGE POINT (1978)
(10^{-9} Ci per kg dry matter per Ci released)

Distance from the discharge point (km) [a]	ΣPu	^{241}Am [b]	^{137}Cs [b]
+0.1	ND	530	54
+1	ND	590	29
+3.3	392 ± 29 (n = 7)	1230	76

Note: ΣPu = ^{238}Pu + ^{239}Pu + ^{240}Pu; ND = Not yet determined; 1 Ci = 3.7 × 10^{10} Bq.

[a] +: Downstream from the discharge point.

[b] Measured by gamma spectrometry (GeLi).

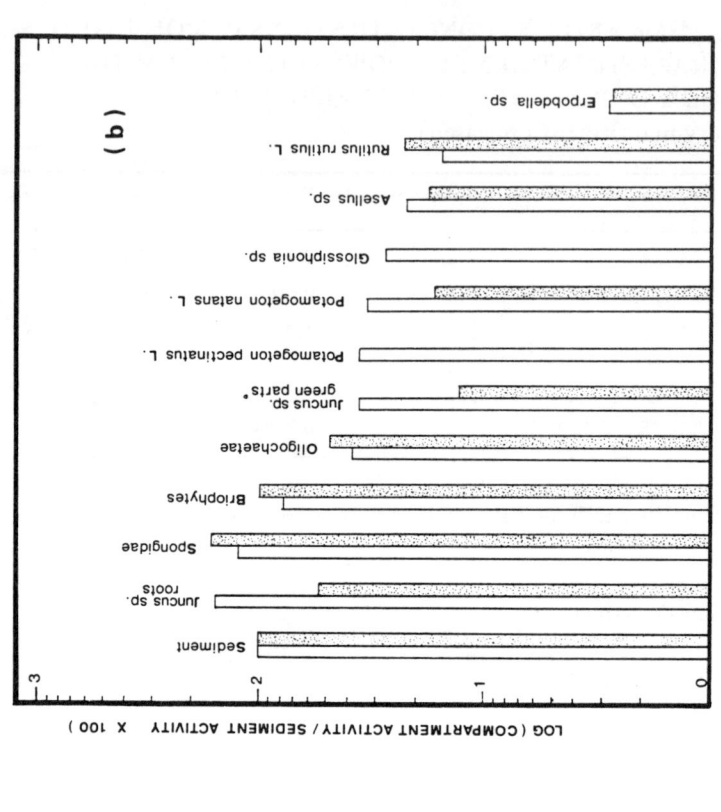

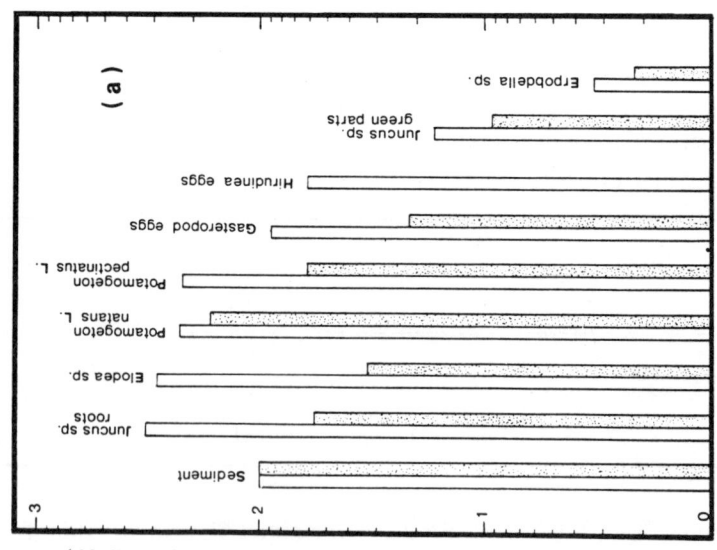

FIG.1. Distribution of ΣPu (□) and ²⁴¹Am (▨) in the Molse Nete ecosystem 1 km (a) and 3.6 km (b) downstream from the discharge point (1978).

TABLE IV. ΣPu, ^{241}Am AND ^{137}Cs CONCENTRATIONS IN MOLSE NETE SEDIMENT CORES (SEDIMENTATION AREA) AS A FUNCTION OF THE DISTANCE FROM THE DISCHARGE POINT (FIRST HALF OF 1978) (10^{-9} Ci per kg dry matter per Ci released)

Distance from the discharge point (km) [a]	ΣPu [b]	^{241}Am [b]	^{137}Cs [b,c]	Pu/Cs	Pu/Am
+1	540	804	54	10.0	0.7
+3.6	1545	1759	261	5.9	0.9
+17.7	15	10	16	0.9	1.5
+35	15	ND	24	0.6	–

Note: ΣPu = ^{238}Pu + ^{239}Pu + ^{240}Pu; ND = Not yet determined; 1 Ci = 3.7 × 10^{10} Bq.

[a] +: Downstream from the discharge point.

[b] $\dfrac{\text{Downstream value} - \text{Upstream value}}{\text{Release}}$

[c] Measured by gamma spectrometry (GeLi).

Data of sediment cores collected in the sedimentation areas at different distances from the discharge point show maximal values 3.6 km downstream (see Table IV), i.e. 1545, 1759 and 261 × 10^{-9} Ci per kg dry matter per Ci released for ΣPu, ^{241}Am and ^{137}Cs, respectively. This was confirmed by another sampling in 1979 with measurements by gamma spectrometry. Apparently this discharge point represents an important sedimentation area. The Pu/Cs ratios diminish with increasing distance from th e discharge point while Pu/Am ratios increase.

Potamogeton natans L. and *P. pectinatus* L. behave in a similar way. When both species are present (see Table V), they have similar ΣPu and ^{241}Am concentrations. *Elodea* sp. attained similar ΣPu and ^{241}Am concentrations as *Potamogeton* spp. The ΣPu and ^{241}Am concentrations in *Juncus* sp. roots are on an average 6 to 7 times higher than in the green parts, an observation which could be explained if the material was adsorbed on the roots. The possibility that sediment had contaminated roots can, however, not be excluded since their cleaning was difficult. *Platyhypnidium* sp., which shows the highest plant concentrations, would make an interesting object as an indicator of actinide concentration.

In general, the concentration of actinides in the fauna (see Table VI) are higher downstream from the discharge point than upstream.

TABLE V. ΣPu AND ^{241}Am CONCENTRATIONS IN MOLSE NETE AQUATIC PLANTS AS A FUNCTION OF THE DISTANCE FROM THE DISCHARGE POINT (1978)
(10^{-9} Ci per kg dry matter per Ci released)

Distance from the discharge point (km)[a]	Potamogeton spp.				Juncus sp.				Elodea sp.		Platyhypnidium riparoides	
	P. natans L.		P. pectinatus L.		Green parts		Roots					
	ΣPu	^{241}Am	ΣPu	^{241}Am	ΣPu	^{241}Am	ΣPu	^{241}Am	ΣPu	^{241}Am	ΣPu	^{241}Am
+0.1	–	–	94±15[b]	–	511±12	500±11	3786±44	3650±57	–	–	–	–
+1	1759±95	455±20	1558±33	266±17	90±2	75±1	1231±50	1330±57	1198±17	489±23	–	–
+3.6	503±12	289±8	–	–	547±10	227±7	2372±42	917±30	–	–	1184±20	1750±30

Note: ΣPu = ^{238}Pu + ^{239}Pu + ^{240}Pu; 1 Ci = 3.7×10^{10} Bq.
[a] +: Downstream from the discharge point.
[b] Mean of 3 counts ± standard error.

TABLE VI. ΣPu AND ^{241}Am CONCENTRATIONS IN MOLSE NETE BENTHIC INVERTEBRATES AS A FUNCTION OF THE DISTANCE FROM THE DISCHARGE POINT (1978)
(10^{-9} Ci per kg dry matter per Ci released)

Distance from the discharge point (km)[a]	*Erpobdella octoculata*		*Glossiphonia* sp.		Hirudinea eggs		Gasteropod eggs	
	ΣPu	^{241}Am	ΣPu	^{241}Am	ΣPu	^{241}Am	ΣPu	^{241}Am
−2.2	–	–	147 ± 34	35 ± 21	–	–	10 ± 8	18 ± 10
+0.1	38 ± 9[b]	74 ± 16	–	–	340 ± 34	235 ± 34	1211 ± 106	1507 ± 146
+1	18 ± 3	18 ± 3	–	–	330 ± 44	–	478 ± 47	172 ± 36
+3.6	43 ± 8	47 ± 13	419 ± 121	–	–	–	–	–

Distance from the discharge point (km)[a]	*Asellus aquaticus* L.		Oligochaeta	
	ΣPu	^{241}Am	ΣPu	^{241}Am
−2.2	5 ± 1	18 ± 3	170 ± 49	–
+0.1	–	–	407 ± 117	–
+3.6	340 ± 18	312 ± 22	595 ± 141	849 ± 206

Note: ΣPu = ^{238}Pu + ^{239}Pu + ^{240}Pu; 1 Ci = 3.7 × 10^{10} Bq.

[a] –: Upstream from the discharge point; +: Downstream from the discharge point;
[b] Mean of 3 counts ± standard error.

The net ΣPu and [241]Am concentrations in *Erpobdella* sp. samples collected 3.6 km downstream represent 2.8 and 1.1% of the sediment activity, respectively, whereas the ΣPu concentration from *Glossiphonia* sp. samples, a species less abundant than *Erpobdella* sp., represents 17.6% of the sediment activity. Oligochaeta display ΣPu and [241]Am concentrations of 27.5 and 55%, respectively, of those in sediment cores, whereas *Asellus* sp. displays concentrations of 21.7 and 19%, respectively, of those of ΣPu and [241]Am in the sediment. Young whole fishes *(Rutilus rutilus* L.), captured 3.6 km downstream from the discharge point, show ΣPu and [241]Am concentrations of 237 ± 4 and 396 ± 6 $\times$ 10^{-9}Ci per kg dry matter per Ci released (based on the releases during the previous 12 months), respectively. Spongidae collected 3.6 km downstream present ΣPu and [241]Am concentrations of 1901 ± 63 and 2867 ± 85 $\times$ 10^{-9} Ci per kg dry matter per Ci released, respectively. From all animal data this group has the highest concentrations.

The distribution of ΣPu and [241]Am at two stations situated downstream from the discharge point is shown in Fig.1, which displays the percentage of activity in a given compartment related to that in the sediment. At low trophic levels highest ratios are found, e.g. in spongidae and bryophytes. The different aquatic macrophytes have similar plutonium ratios *(P. natans* L., *P. pectinatus* L.- *Elodea* sp.).

Roots of plants emerging from the water surface *(Juncus* sp.) show higher ratios than the green parts. The plutonium ratios in oligochaetae are of the same order of magnitude as in aquatic macrophytes *(Potamogeton)* but the americium ratio is higher. Fishes *(Rutilus,* whole) represent the highest trophic level in this fresh-water ecosystem and have a low plutonium ratio which is in agreement with literature data.

REFERENCES

[1]   VAN DE VOORDE, De rol van de riviersedimenten bij de rekoncentratie van effluent radioaktiviteit, Thesis,;, UCL, Louvain, Belgium (1968).

[2]   SAAS, Sites de la Witte Nete et de la Molse Nete. I. Caractéristique des eaux et des sédiments, CEN Cadarache (1978).

[3]   SINGH, H., MARSHALL, J.C., A preliminary assessment of [239,240]Pu concentrations in a stream near Argonne National Laboratory, Health Phys. **32** (1977) 195.

[4]   BARTELT, G.E., et al., Plutonium concentration in waters and suspended sediment from the Miami River Watershed, Ohio, Argonne National Laboratory, 75-3, 72-77.

SUMMARY AND RECOMMENDATIONS

1. INTRODUCTION

The papers presented in the biological section ranged from freshwater to marine studies, and from laboratory to experimentally derived data. A large number of relevant remarks were made in all of these, many of which were interrelated. Thus, instead of summarizing each paper individually, the relevant points of emphasis have been discussed collectively by the group.

2. DISCUSSION

It is clear that although there are various sources of transuranium nuclides such as world-wide fallout, the Pacific test sites, effluents from reprocessing plants and so on, the state of one element, plutonium, in filtrate sea water, appears to be the same. It is possible that other transuranium elements may also behave similarly; if not immediately, then after a certain period of time. Unfortunately, the spread of values obtained in biological materials from different sites is such that, at present, overall trends on the importance of different source terms are difficult to discern. It is also true that data are principally available for plutonium, and there are so few data on the other transuranium nuclides that generalizations are impossible. This general remark applies to the whole report of the biological working group.

As far as individual organisms are concerned, the important aspect is the physico-chemical partitioning which occurs external to the organism, i.e. the partitioning between water, sedimentary materials, and the organism's food; and this, in turn, is dependent upon the speciation of the nuclide in the immediate environment. For example, if the radionuclide is in the III oxidation state, such as americium or curium, then it is to be expected that such a nuclide will probably be adsorbed onto algal surfaces, onto animals with a high surface-to-volume ratio, and be ingested by particulate feeders. In contrast, those which remain fairly soluble may be more biologically available to pelagic organisms. It is also important to recognize that, although perhaps accumulated whilst adsorbed onto particulate material, the nuclide may also be internally solubilized and translocated within an organism from one organ to another. Yet a further complication is the effect of parent-daughter relationships, so that a radionuclide may appear in an organ as the result of grow-in — such as ^{241}Am from ^{241}Pu, and ^{238}Pu from ^{242}Cm.

Although concentration factor data are usually sought, because of the rapid removal of many of the transuranium nuclides to sediments, and because of

rapid changes in the ambient water levels, other data are required to characterize the extent of bioaccumulation. More experimental data are required on
the transfer of such nuclides to organisms from contaminated sediments and
from different forms of food species.

It is recommended that in all such experiments — be they transfer from
water, sediments, or food — great care should be taken to control the initial
chemical form used (e.g. Pu(IV) or (VI)), and to observe whether these chemical
states change during the course of the experiment. Other water qualities should
also, of course, be carefully monitored in experimental systems and the physiological state of the organism should also be carefully considered. It is also
important to remember that, in any extrapolation of derived data to environmental conditions, differences may occur because of the frequently large
differences in molar concentrations which are used. The use of recently available
isotopes, such as ^{237}Pu, is of particular value in such studies. In many cases it
is, therefore, not possible to derive absolute concentration factor data. Nevertheless, some useful data can be obtained by making observations on relative
concentrations of different nuclides in the organism and in other surrounding
materials — such as those in the filtrate and particulate fractions of the water —,
in the sediment, in other species — such as food species — and in the source
term. It is also important to note that concentration factors only relate to
steady-state conditions and, in many cases, these are unlikely to obtain. The
transfer factors derived from the above-mentioned approaches should be used
to obtain systems analysis models of radionuclide transfer through different
links of the food chain, relative to the source.

In order to generalize on the metabolism of transuranium nuclides by
aquatic organisms, basic metabolic studies, for example subcellular studies, are
very useful — although by no means essential from the radiological protection
point of view — in order to see if there is any commonality in the observations
which are made. Also of value are comparisons with naturally-occurring
analogues — such as uranium and thorium for the different oxidation states of
plutonium, neodymium for americium and so on — although great care is
required, and considerable caution is urged, in drawing too many conclusions
from such observations. These data will only indicate possible common trends,
but may nevertheless be very useful as indicators of the long-term distribution
and biological availability of transuranium nuclides after they have been cycled
many times through the biosphere. Measurements of naturally occurring alpha
emitters in general, and of ^{210}Po in particular, are also most useful in keeping
the impact of transuranium nuclides in aquatic organisms, as a source of radiation
exposure to man, in perspective.

In comparing different aquatic environments, the majority of data, so far
available, are those arising from marine waters and the Great Lakes. There are
too few data on other aquatic environments. It is also important to realize that

within any one general environment different environmental regimes may prevail, which locally may be very important. As well as studying different environments in situ it is recommended that hypotheses be tested under controlled laboratory conditions. This would be of particular value using brackish water organisms for example because the same species can be used over a wide range of environmental conditions. Species with a wide ecological range should also be used for intercomparability, and to validate their usefulness as indicator organisms. Theoretical methods are becoming available which may make it possible to calculate the distributions of transuranium nuclides with different ligands in different aquatic regimes.

In any aquatic environment, transfer along food chains is of great interest. This information will only be of value, however, if, in reporting such data, it is explicitly stated whether comparisons are made on the basis of whole-body, or selective organ, analyses.

Laboratory experiments are again of value here, particularly to find out if the effect of transfer through the food chain actually alters the subsequent biological availability of the nuclide.

Another interesting approach, and one which is of particular value in radiological protection, is to relate the concentrations in any link in the food chain, to the rate of radionuclide released from the source: for example as $fCi \cdot g^{-1}$ per $Ci \cdot d^{-1}$ released. This approach has been successfully applied to other radionuclides.

Bacterial action is often implicated as a mechanism for altering the biological availability of radionuclides in the aquatic environment, but there is a lack of data on this subject. It is recommended that, initially, simple experiments be made. Environmental cores used for interstitial water characterization should also be characterized with regard to the degree of bacterial activity within the core profile.

The effects of bioturbation on radionuclide distributions within sediments is another area which requires more detailed examination. The interpretation of environmental materials would be greatly enhanced if careful records were made, for each core, of the organisms found within them; and, ideally, cores should be X-rayed on board ship in order to provide a record of the physical evidence of biological disturbance before the core is sectioned for analysis. Laboratory experiments could also be made — using layered sediments which are appropriately labelled — to determine more precisely the mechanical effects of biological action. Such experiments could be combined with chemical speciation analyses of the radionuclides which are introduced and recovered from the experimental system.

Radiation effects of transuranium nuclides have not been addressed at this meeting, but it is suggested that as further data become available the present dose models, and estimates of absorbed dose, be considered by the relevant experts in this field.

Finally, the effects of other contaminants in aquatic environments must be considered. It is to be expected that the presence of such contaminants will play an important role in altering the relative biological availability of trans-uranium nuclides.

3.   RECOMMENDATIONS

(1) In order to determine if generalizations can be made on the bio-availability of all transuranium nuclides in aquatic ecosystems, more data are required from different environments and in areas related to different source terms. The use of the same, or similar, species at different locations is of particular interest in this respect. In comparing different environments, more information is required on the statistical distribution of the data. This, of course, is impractical for transuranium nuclides because of the very low levels encountered, which usually requires that samples be bulked for analysis; but data on other, more readily measurable nuclides, could provide useful indications.

Isotope ratios are also useful in comparing different environments in rela-tion to different source terms, and particularly in comparing the relative biological availability of plutonium, americium, curium and neptunium.

(2) More data are required on the transfer of transuranium nuclides to organisms along pathways other than direct uptake from water, i.e. from contaminated sediments and food.

(3) Greater use should be made of naturally-occurring analogues for comparison with transuranium nuclides. Of course, such data will only provide useful directions for further research, and cannot necessarily be directly extra-polated to the transuranics.

(4) Microbiological studies are required to estimate the importance of this component of the aquatic environment in affecting the bioavailability of transuranium nuclides. More data from environmental cores are required to complement chemical studies on the interstitial regime. Suitable laboratory studies, along the lines indicated in the discussion, should also be considered.

(5) The role of bioturbation in altering the physical and chemical distri-bution of transuranium nuclides in sediments needs to be studied in more detail, both from the point of view of environmental observations on cores which are taken, and from suitable laboratory studies discussed above.

(6) The effects of other contaminants in altering bioavailability should be studied in more detail, and advantage taken of such areas where these contaminants are known to be introduced.

Session III

CHEMICAL METHODS
OF DETERMINING METAL FRACTIONS
AND APPLICABILITY TO TRANSURANICS
ON FRESH, ESTUARINE AND COASTAL SEDIMENTS

THE PARTITIONING OF TRACE METALS
AND TRANSURANICS IN SEDIMENTS

R. CHESTER
Department of Oceanography,
University of Liverpool,
Liverpool

S.R. ASTON
Department of Environmental Sciences,
University of Lancaster,
Lancaster,
United Kingdom

Abstract

THE PARTITIONING OF TRACE METALS AND TRANSURANICS IN SEDIMENTS.
    This paper suggests chemical separation techniques that have been useful for trace
element investigations as a means to acquire reliable data on the partitioning of transuranics
in sediments. A caution is noted that no single technique will identify all specific transuranic
associations within the non-residual fractions of the sediments. A sequential leaching scheme
is shown and can be adapted to additions at various stages of separation to gain information
on transuranic associations which were not included in the basic scheme. The importance of
pollutant distribution patterns based on relative extraction methods is illustrated.

1. INTRODUCTION

On a historical time-scale sediments may be regarded as at least a temporary
sink for much of the material which passes through the various aquatic chemical
and biological cycles operative on the earth's surface. At present, the material
which is released to take part in these cycles can have its composition markedly
affected by anthropogenic emissions, and sediments therefore become an environmental host for many of the waste products discarded by society. The effects of
these man-made emissions will vary greatly from one place to another, but in
some situations they can be of sufficient strength to impose 'anthropogenic
fingerprints' on the deposited sediments.

Certain sediments are at present accumulating at such slow rates that the
fingerprints from anthropogenic emissions cannot yet be detected on a gross scale.

173

Perhaps the prime examples of such sediments are those which are forming in the deep-sea regions at rates which are of the order of $mm/10^{-3}$ a. That is, over the last millenium only ~ 1 mm of sediment has been deposited in these remote pelagic areas, and since the principal anthropogenic emissions have occurred over the last few hundred years it is apparent that even if the upper few millimetres of the sediment column could be sampled accurately it would not yet be possible to fully distinguish the effects of these emissions. Nonetheless, even in these remote regions anthropogenic substances such as radio-isotopes (see e.g. Ref. [1]), synthetic high molecular weight organic chemicals (see e.g. Ref. [2]) and elemental carbon (see e.g. Ref. [3]) have been identified in surface sediments, or in traps placed above them. However, in the marine environment it is near-shore sediments (i.e. those deposited in the coastal, lagoonal and estuarine regions) which will retain the strongest memory of the impact of the release of anthropogenic material onto the earth's surface. There are a number of reasons for this, two of the most important being: (a) these sediments are accumulating at rates which can be as much as several orders of magnitude higher than those operative in the deep sea, with the result that recent changes in input parameters can be recorded, and (b) they are forming in the very areas in which many anthropogenic substances initially reach the seas, i.e. at the boundaries between the continents and the oceans. On the continental slope itself a variety of sediments, e.g. those of the lacustrine, pond and reservoir environments, can reflect anthropogenic emissions. Over the past decade or so it has become increasingly apparent that some near-shore and continental sediments of these various types can, and indeed do, retain anthropogenic fingerprints.

Once a substance — and the present review concentrates on trace metals and transuranics — has been incorporated into a sediment, its ultimate fate depends on a number of very complex factors. Expressed in the simplest terms, a trace metal or transuranic may be considered to be locked 'permanently' into a sedimentary component, or it may subsequently be released and take part in various bio-geochemical reactions, some of which may be environmentally hazardous. The ability to predict the subsequent fate of trace metals and transuranics incorporated into sediments is therefore one of the key factors in the assessment of the effects of environmental pollution. In order to be able to make even the most primitive predictions of this kind, it is necessary to understand something of the mechanisms by which these elements are incorporated into sediments. It is no longer sufficient simply to carry out chemical analyses on the total samples of a sediment population and to attempt to deduce from them the degree of pollution to which the deposits have been subjected. Rather, it is the processes of trace metal or transuranic retention and release that must be delineated.

One of the fundamental distinctions that has long been made in sedimentary geochemistry is that between material brought to the site of deposition in a solid form, and that brought in a dissolved (or colloidal) form. Krynine [4] used this

simple concept of solid and dissolved material to make a two-fold classification
of sediment components into a detrital fraction (transported as solid material), and
a non-detrital fraction (transported as dissolved or colloidal material). However,
although it is theoretically useful, this approach has severe limitations when applied
to the study of the partitioning of trace metals and transuranics in sediments.
For example, metals located in the non-detrital fraction of a sediment have not
necessarily been removed from the overlying waters at the present site of deposition,
but could have been transported from elsewhere in association with non-detrital
components. Nonetheless, the concept of distinguishing between solid and
dissolved material provides a basic framework within which to discuss the origins
of the components of sediments, and it has been extended by various authors
into more sophisticated classification schemes in which the non-detrital fraction
has been sub-divided into several individual genetically separated fractions
(see e.g. Refs [5–9]).

Another disadvantage of classifying elements into either detrital or non-
detrital types is that under certain circumstances it may be extremely difficult
to distinguish between the two on practical grounds, due to either theoretical
and/or analytical considerations. Sediment components which pose analytical
difficulties include opaline silica, which is extremely difficult to separate chemically
from detrital alumino-silicates. An example of the limitations imposed by
theoretical considerations is provided by those elements which have a non-detrital
origin but which may subsequently be incorporated into detrital components.
This can occur during the formation of some secondary hydrogenous minerals
(see e.g. Ref. [6]), such as the zeolites, when metals removed from sea water may
be taken into the mineral lattices (see e.g. Ref. [10]).

The transuranic elements have no stable counterparts in the natural environment, but for certain other radionuclides isotopic exchange with stable forms can
be important. In this context it is worth noting that radionuclides have provided
direct evidence of the incorporation of anthropogenic materials into detrital phases
in sediments. For example, ^{137}Cs has been found to diffuse into the lattices of
clay minerals after adsorption [11], although diffusion of this kind is characterized
by very low diffusion coefficients [12]. Replacement of the ions of other elements in
the lattices of detrital minerals is another possibility which must be considered
for radionuclides, including the transuranics. The particular importance of the K^+
ion in ^{137}Cs and [134]Cs adsorption into sediments is well established, and accounts
for the particular association of caesium isotopes with the illite fraction of
sediments (see e.g. Refs [11, 13–16]). This type of adsorption reaction has also
been found in soils and freshwater sediments (see e.g. Refs [17–19]). Such
replacements of lattice ions are probably responsible for the lack of desorption
of some pollutants, e.g. caesium isotopes, from sediments [20, 21]. These detrital
mineral substitution reactions have not as yet been reported for the transuranic

elements, but the possibility that they occur does exist, and has quite important implications to their long-term retention in sediments.

For reasons such as those given above it is therefore advantageous to make an even more fundamental distinction than the one between the detrital and the non-detrital fractions of sediments, and that is to characterize a lattice-held (i.e. residual) and a non-lattice-held (i.e. non-residual) fraction. This distinction is very important in respect of the subsequent fate of trace metals and transuranics incorporated into sediments, whether they have a natural or an anthropogenic origin. The rationale underlying this is that trace metals located in lattice positions can usually be considered, at least to a first approximation, to be immobile, i.e. environmentally unreactive, whereas those in non-lattice sites can be considered to be at least potentially mobile, i.e. environmentally reactive, in the chemical and biological processes which occur in the sediment/ interstitial water complex. This classification of elements into *residual* and *non-residual* types is particularly appropriate to the investigation of trace metal and transuranic pollution because the processes (e.g. adsorption, complexation) which are principally involved in the incorporation of these elements from polluted waters result in their being located in the non-residual fractions of sediments. Because of this, the analysis of the non-residual fractions will often provide more data on the extent to which a sediment has suffered from the effects of pollution than will a total sediment analysis, since the latter includes the residual, or non-polluted, fraction which may mask the relationships sought (see e.g. Refs [22, 23]).

The transuranics have no stable counterparts in the environment and thus they are not present in the residual fraction of sediments, unless some (as yet not observed) diffusion into detrital minerals occurs (see above). Transuranic pollutants will be present, to greater or lesser extents, in the sedimentary components which make up the host materials of the non-residual fraction. It is, therefore, apparent that the application of chemical partitioning techniques will not be of great interest for the transuranics when those techniques are limited to a simple residual and non-residual classification. However, the more sophisticated chemical partitioning of the non-residual fraction into its constituent phases may provide fundamental information on the sedimentary geochemistry of these elements. Nonetheless, the present review is concerned with both stable trace metals and transuranics, and for this reason the usefulness of a variety of partitioning techniques is considered; and these include techniques designed to distinguish only between the residual and non-residual sediment fractions.

In general, elemental partitioning studies in sediments should be directed towards providing information on the following: (a) the identification of pollution; (b) the extent of the pollution; (c) the origins of the pollutants; and (d) their subsequent fate. With information such as this it should be possible to achieve some understanding of the nature of the anthropogenic fingerprints which man is imposing on the environment, and thus be a little closer to attaining the overall aim, i.e. the management of the global environment.

2.   TECHNIQUES FOR THE INVESTIGATION OF TRACE METAL PARTITIONING IN SEDIMENTS

In general, there are three principal ways in which the partitioning of trace metals and transuranics between the various sedimentary components can be studied. These are: (a) by the interpretation of the chemical analyses of total sediment samples, (b) by the physical separation and subsequent analysis of individual components, and (c) by the chemical separation and subsequent analysis of the components. Each of these is considered individually in the following sections.

2.1. Interpretation of total sediment analysis

By definition, total sediment samples include both the residual and the non-residual fractions. One immediate problem which arises therefore in the interpretation of total sediment analyses is that the residual trace metals will serve to dilute, and so may mask the importance of the non-residual metals. In practice, a great deal of information on the partitioning of trace metals and transuranics in sediments can be obtained from the total sample analysis. However, such analyses need very careful interpretation.

In order to identify the presence, or absence, of anomalous trace metal concentrations in sediments it is necessary to establish the natural, or baseline, concentrations in unpolluted deposits; a consideration which is not necessary for the transuranics. However, there are considerable difficulties inherent in the selection of a coastal sediment baseline material, many of which arise because of the very wide variety of environments under which such deposits may be formed. For example, redox conditions, which can exert a considerable influence on the geochemistry of a sediment, vary greatly from place to place. Förstner and Wittmann [22] have considered the problem of the selection of a sediment baseline material for environmental studies and have suggested a number of criteria that should be fulfilled in order to best satisfy the principal requirements. The criteria which the baseline material should satisfy include the following: (a) its conditions of origin, grain size and material composition should correspond to those of recent deposits; (b) it should be uncontaminated by anthropogenic influences; and, (c) the analyses of a large number of samples should be available. In practice, all of these requirements are almost impossible to fulfil simultaneously, but as Förstner and Wittmann [22] have found out, fossil argillaceous sediments, i.e. shales, probably offer the best medium against which to compare the compositions of present-day aquatic sediments, at least on a general level.

Once a background material has been selected, there are still problems with regard to its direct application to a particular suite of sediments. One of the most important of these problems is the effect on the composition of the sediments of components which are impoverished with respect to the elements under consideration.

For most trace metals calcium carbonate shell material is a prime example of such a component, and in the total sediment it will act as a diluent on the metals present in the non-carbonate fractions. The effects of this dilution can seriously hinder the interpretation of total sediment analyses because the concentrations of calcium carbonate can vary considerably from sample to sample over relatively small areas. There are a number of ways of overcoming this difficulty, most of which involve some form of recalculation of the compositional data. The most simple approach is to express the elemental analyses on a carbonate-free basis, although this has several drawbacks. For example, for those sediments which are rich in carbonate, even small errors in the estimation of this component will severely affect the recalculated elemental concentrations and, since many coastal sediments contain $\gtrsim 75\%$ $CaCO_3$, this can be important. Further, expressing the analyses on a carbonate-free basis takes no account of the diluting effects of other trace metal-poor components, such as opaline silica. Problems such as this have led to more sophisticated recalculation schemes in which the total sediment analyses are expressed in terms of a silicate component [24], an abiogenic component (see e.g. Ref. [25]), or a minerogen component.

Another approach which may be used for the comparison of sediments of different compositions to a single baseline involves the use of specific element as an indicator of a general source material. In this approach it is usual to employ an enrichment factor (EF) which is calculated in the following manner;

$$EF_{(X)} = \frac{(\text{Conc. X/Conc. Ref.}) \text{ in analysis material}}{(\text{Conc. X/Conc. Ref.}) \text{ in source material}}$$

where $EF_{(X)}$ is the enrichment factor of an element X in the analysis material relative to a reference element (Ref.) and in a source material relative to the reference element. For sediments, the use of indicator elements can be taken a stage further, and rather than indicating a general elemental source, they can be used to establish the location of the elements in specific sediment fractions. In terms of distinguishing between the location of an element in either the *residual* or the *non-residual* fraction of sediments, the most commonly used indicator element is aluminium. The rationale underlying this is that in most sediments, i.e. those which do not suffer direct aluminium pollution, this element is almost exclusively located in the lattice structures of alumino-silicate minerals; that is, it is almost exclusively residual in character. By the use of enrichment factors relative to aluminium, or simple element/aluminium ratios, it is often possible to estimate the partitioning of an element between the residual and non-residual fractions of sediments providing the enrichment factors, or ratios, are known for aluminosilicate residual minerals. In practice, common rock types for which many analyses are available are used as baseline materials, rather than individual minerals. It was suggested above that shales offer the best baseline against which to relate

TABLE I. AVERAGE PARTIAL ELEMENTAL COMPOSITION OF SHALE AND A SEDIMENT SAMPLE FROM THE MERSEY RIVER (UK)

Element	1. Shale		2. Mersey mud	
	Concentration (ppm)	Element/Al ratio $\times 10^3$	Concentration (ppm)	Element/Al ratio $\times 10^3$
Al	100 000	—	57 460	—
Fe	47 200	472	28 540	497
Mn	850	8.5	1 903	33
Cr	90	0.9	58	1.0
V	130	1.3	—	—
Ni	68	0.68	39	0.68
Co	19	0.19	16	0.28
Cu	45	0.45	212	3.7
Pb	20	0.20	187	3.3
Zn	95	0.95	829	14
Ba	580	—	—	—
Sr	300	—	—	—
As	13	—	—	—
U	3.7	—	—	—
Be	3.0	—	—	—
Sb	1.5	—	—	—
Hg	0.4	—	—	—
Cd	0.13	—	—	—

the compositions of present-day aquatic sediments. Although shales are 'natural' deposits, i.e. they have not been subjected to pollution, they do contain both a residual and a non-residual fraction. However, it can be assumed, at least to a first approximation, that because of the residual nature of aluminium, anomalously high element/aluminium ratios or enrichment factors, in present-day sediments relative to shales are a result of anomalies in their non-residual fractions. Further, the use of the ratios, or enrichment factors, does allow a direct comparison to be made between the shale baseline material and coastal sediments of any mineral composition, because these ratios will be independent of the presence of components such as calcium carbonate and opaline silica which do not contain aluminium.

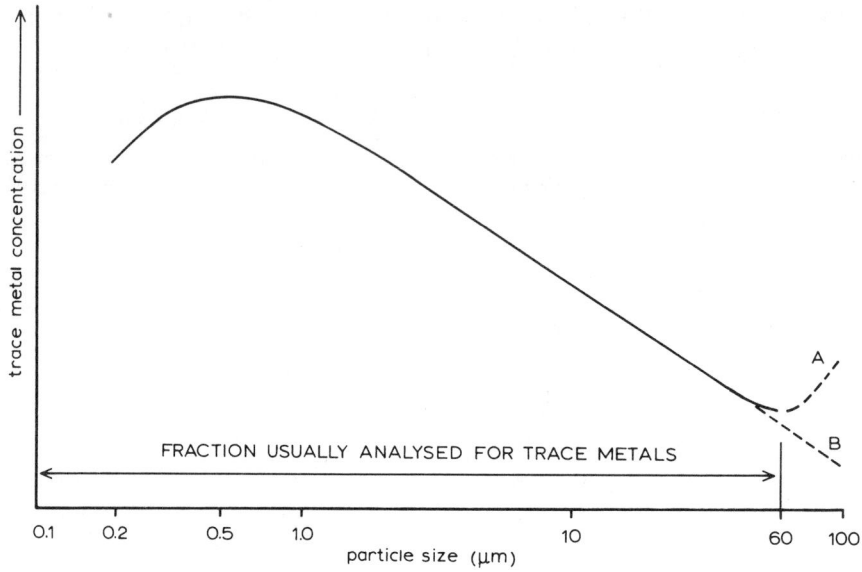

FIG.1. A generalized profile of the variation of the concentration of a trace metal with grain size in a sediment. (Modified from data given by Förstner and Wittmann [22].)

The average partial element composition of the shale baseline material is given in Table I, together with the trace metal composition of a river sediment. Trace metal/aluminium ratios are also included in this table, and it is evident that those of manganese, copper, lead and zinc are all at least one order of magnitude higher in the river sediment than in the shale; i.e. it is for these elements that a pollutant source must be suspected.

2.2. The physical separation of sediment components

The physical separation of individual sediment fractions from most present-day sediments, particularly muds, is time consuming, difficult and, if used for chemical analysis, open to contamination. However, there is one application of the physical separation of sediment fractions that is extremely important, and that is in respect of grain size. In general, there is an overall increase in trace metal concentrations with decreasing grain size in many sediments. A generalized profile of the variation of a trace metal with grain size in a sediment is given in Fig. 1. Several points emerge from a study of this type of diagram: (a) between grain sizes of ~ 0.2 and ~ 60 μm there is a general increase in trace metal concentration with decreasing grain size; (b) at grain sizes of $\lesssim 0.2$ μm there is often a decrease in trace metal concentration which may be due to a reduced absorption potential of less crystalline, or amorphous, material [22]; (c) at grain sizes of $\gtrsim 60 \, \mu m$ one of two general conditions may apply; the concentration curve may be raised as a result of the presence of trace metal-rich heavy minerals (curve A), or it may be lowered due to the presence of large-sized trace metal-poor minerals such as quartz (curve B).

Little work has been carried out on the distribution of transuranics in sediments with regard to grain size characteristics, especially in the $\lesssim 60 \, \mu m$ fraction in which important relationships may be expected to occur. For the transuranic nuclides a relationship between concentration and grain size distribution has been found both in the Windscale (UK) area [26–28] and in the Buzzards Bay area [29]. In estuarine sediments near Windscale, the α-emitting contaminants among the various sizes of particles have shown a steady increase in the specific activity going from sand towards clay. This leads to the same conclusions found for the fission products, i.e. that the uptake is essentially a surface adsorption phenomenon with each of the phases of the sediment reconcentrating the activity to a degree determined by its surface area. However, the evidence from other regions show that this is not always the case. For example, Bowen et al. [30] found no strong preference of $^{239, \, 240}$Pu for the finer grain sizes; this is similar to results found in freshwater lakes (see e.g. Ref. [31]).

In order to reduce the effects on trace metal or transuranic concentrations which result from grain size differences, it is common practice to separate a particular sediment fraction, e.g. by wet sieving, and so allow the direct intercomparison of data obtained from sediments collected from a wide variety of depositional environments. In most techniques, the fraction which has a grain size of $\lesssim 60 \, \mu m$ is used for analysis.

2.3. The chemical separation of sediment components

Potentially, the chemical separation of components offers the most scientifically satisfying approach to the study of elemental partitioning in sediments because the processes by which the metals are incorporated into the various components are themselves largely chemical in nature. In theory at least, it should therefore be possible to 'unlock' these processes, and over the past decade or so a great deal of research has been directed towards this end. However, the sedimentary medium is complex and although scientists do have some insight into its nature, the study of the partitioning of trace metals and transuranics in sediments must still be cautiously regarded as 'the art of the possible'.

It is convenient to discuss chemical separation techniques in the context of the sediment fraction, or fractions, for which information is required. This is important since most techniques employing a chemical attack on a sediment involve a leaching procedure which is specifically designed to put into solution part, or sometimes all, of the total sediment sample.

Total sediment dissolution. In some forms of analysis, e.g. atomic absorption spectrophotometry and alpha spectroscopy, it is necessary to bring all the sediment sample into solution. In other trace metal analysis techniques, e.g. neutron activation analysis, samples can be presented to the instruments as solids, and it is only necessary to use leaching techniques for the separation of particular sediment fractions. In those techniques for which a solubilization of the total sediment is required it is necessary to completely break down the silicate lattice material. This is usually accomplished by attacking the sample with HF, in combination with some other concentrated acid, or acids, such as HCl, HNO_3 or $HClO_3$.

Selective sediment dissolution. It was shown above that it is useful to make a fundamental distinction between residual (i.e. those associated with the sediment matrix) and non-residual elements (i.e. those which have been incorporated into the sediment from aqueous solution), and there are numerous chemical techniques which have been designed to effect the separation of these two fractions. The techniques vary in complexity and may be divided into those which identify the total non-residual elements and those which isolate elemental associations within the non-residual fraction.

One of the most simple chemical techniques which may be used to separate the total non-residual from the residual trace metals involves leaching the sediment with a cold weak acid (see e.g. Ref. [23, 32]). A simple, rapid and reproducible cold acid leaching technique of this type has been applied to a series of Greek coastal sediments by Chester and Voutsinou [33], and the results will serve to illustrate the usefulness of such a technique for the initial identification of trace metal pollution in estuarine and coastal deposits.

In the technique used by Chester and Voutsinou [33], the $< 62 \mu m$ fractions of a series of total sediment samples were shaken for ~ 16 hours with cold 0.5N HCl, filtered, and the filtrate sprayed directly into the atomic absorption instrument. The samples were collected from two Greek coastal gulfs: Thermaikos Gulf, for which the sediments were expected to show the effects of trace metal pollution, and Pagassitikos Gulf, which was considered to be relatively 'unpolluted'.

There are a number of ways in which the data obtained from a survey of the distribution of non-residual elements can be interpreted. Three of the most commonly used of these involve: (a) the evaluation of spatial elemental variations, (b) the comparison of elemental concentrations in sediments from polluted and non-polluted regions, and (c) the evaluation of vertical, i.e. time-dependent, elemental variations. Each of these is considered below.

(a) Spatial elemental variations in surface sediments sometimes offer a dramatic manifestation of the effect of pollution because, under certain circumstances, they can almost directly pinpoint the principal source, or sources, of the anthropogenic emissions. An example of this is provided by the distributions of non-residual trace metals in the Gulf of Thermaikos, and even from such an initial

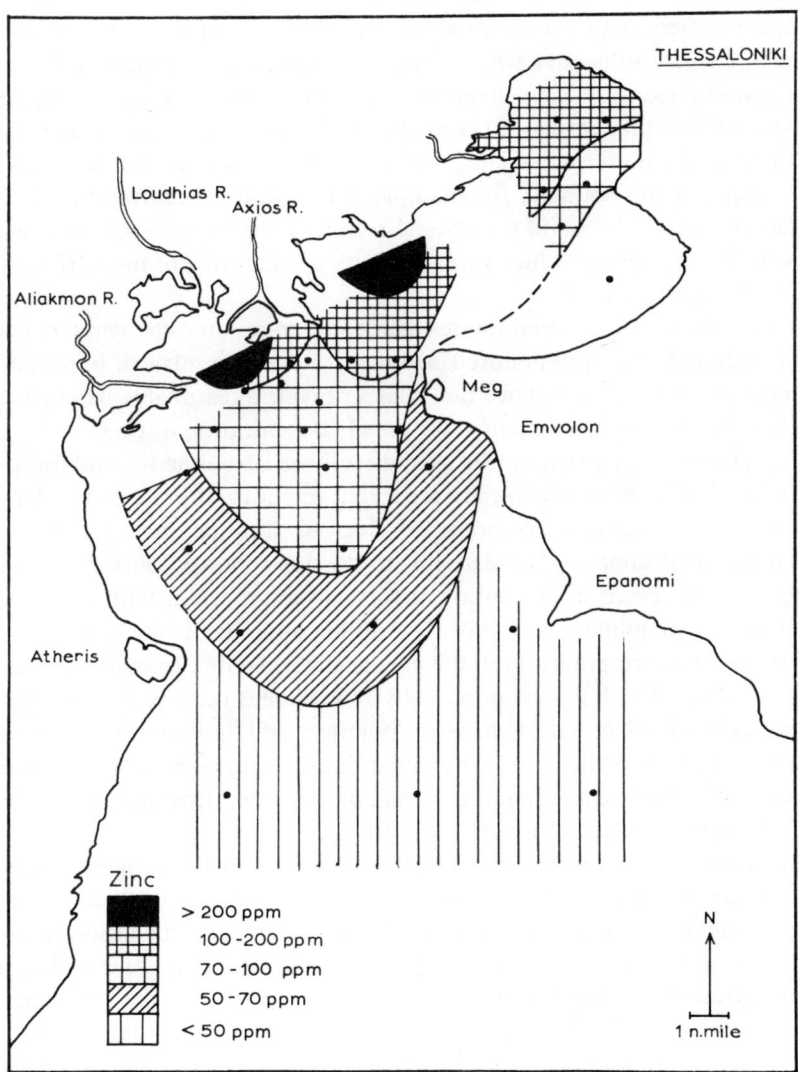

FIG.2. The distribution of non-residual zinc in surface sediments from Thermaikos Gulf (Greece) [33].

survey it was apparent that for some elements there were distinct 'anthropogenic fingerprints' in the sediments. To illustrate this, the distribution of the total non-residual zinc in the surface sediments is shown in Fig. 2. In this figure the zinc concentrations have been contoured, and although this is a crude procedure which is extremely subjective when carried out manually on relatively few samples, it does permit regions of 'high' elemental concentrations to be identified; regions from which the sediments can subsequently be subjected to more detailed analysis. Contouring of the non-residual elements in this manner can also provide information on the anthropogenic sources; for example, the highest concentrations of zinc in Thermaikos Gulf are found in the general region of the outflows of the Loudhias and Axios Rivers, both of which receive inputs from an area of industrialization before reaching the gulf.

(b) Comparison between the residual and non-residual elemental concentrations of 'polluted' and 'non-polluted' sediments raises a number of important problems. It was suggested above that shales provide a reasonable baseline against which to relate the *total* sediment trace metal concentration of present-day aquatic deposits. However, shale cannot be used directly as a baseline to establish the degree of pollution in the chemically separated fractions of present-day deposits since the average trace metal composition given for it in the literature refers to the total sediment sample. To relate this to the non-residual fractions of other sediments it is necessary to use total sample analyses, together with an indicator element such as aluminium (see above). Another approach is therefore required to assess the degree of pollution in the chemically separated fractions of recent deposits, and one which is least potentially rewarding is the direct comparison of the non-residual trace metal fractions of 'polluted' and 'non-polluted' sediment populations, providing that the two populations have reasonably similar conditions of deposition, grain size and major mineral composition. However, the selection of such baseline populations is very difficult.

One possible way of obtaining baseline sediments of this kind is to select them from particular parts of the region in which the non-residual elemental survey is being made. For example, it is evident from the spatial distribution of non-residual zinc in the surface sediments of Thermaikos Gulf (Fig. 2) that, in general, the concentrations of this element decrease markedly towards the southern portions of the gulf. Further, in the most southerly sediments there are no discernible patterns in the distribution of zinc. It may be possible, therefore, to use these sediments as a baseline against which to assess the extent of the non-residual trace metal pollution in the samples from the northern gulf. The average non-residual concentrations of zinc and lead, in the northern and southern sediment populations are given in Table II, from which it can be seen that the northern deposits are indeed relatively enriched in both of these metals. However, evidence of this kind is not, in itself, sufficient to characterize the southern sediments as being non-polluted 'background' material because anthropogenic

TABLE II.   AVERAGE NON-RESIDUAL CONCENTRATIONS OF ZINC AND
LEAD IN SURFACE SEDIMENTS FROM TWO GREEK GULFS[a]

Element	Thermaikos Gulf; northern section Conc. (ppm)	Thermaikos Gulf; southern section Conc. (ppm)	Pagasiticos Gulf Conc. (ppm)
Zn	110	44	21
Pb	83	33	20

[a] Data from Chester and Voutsinou [33].

emissions may still play a part, albeit a smaller one, in controlling the level of the
non-residual trace metal concentrations in them. In order to overcome this
difficulty, sediments from a relatively non-polluted region, i.e. Pagassitikos Gulf,
were also analysed, and the average non-residual concentrations of zinc and lead
in them are given in Table II. It is apparent from this table that the non-residual
concentrations of zinc and lead are both lower in the Pagassitikos sediments than in
those from the southern region of Thermaikos Gulf; thus, highlighting the
difficulties of selecting background sediments, particularly for the assessment of
non-residual trace metal pollution. It is tempting to conclude, therefore, that in
the investigation of non-residual elemental pollution, spatial surface contouring
of the metal concentrations offers the most readily interpretable data. Background
concentration levels are most certainly needed, but it is evident that great care
must be taken in choosing the best sediment population from which to establish
such levels.

　　　For the transuranic elements, the question of the selection of background
concentrations does not arise in the same sense as it does for the trace metals.
Fallout of plutonium isotopes from atmospheric nuclear weapons' tests does,
however, provide a rather different 'background' concentration in sediments. This
anthropogenic 'background' may be used in the assessment of the contribution
of plutonium to sediments from other sources, e.g. nuclear fuel reprocessing
wastes.

　　　(c) Down-core, or time-dependent, trace metal and transuranic variations
can often yield valuable environmental data; however, the interpretation of such
data must be treated with extreme caution. In the most simple situation, a
sediment which has received pollutant emissions can be expected to record these
emissions as a function of time, and with accurate dating it should be possible

to pinpoint at least the general period in which the anthropogenic releases were initiated. In nature, however, this simple relationship can be strongly modified by several processes; two of the most important of which are post-depositional migration and sediment mixing by burrowing organisms.

The post-depositional migration of trace metals in sediments has been discussed by many authors — for a review of the subject see Price [34]. In many ways it is difficult to assess the importance of post-depositional migration because it is difficult to separate its effects from those of variations in anthropogenic and natural emission levels. An insight into this problem may be gained from a study of those radionuclides which are exclusively man-made, since at least their down-core profiles cannot be complicated by the upward migration of elements which were incorporated into the sediments before the onset of any anthropogenic emissions.

There have been several studies on the down-core variations in transuranic concentrations, especially in the near-shore marine and lake environments. In particular, there are two contrasting sets of observations which are, however, not wholly contradictory, as they apply to rather different sedimentary regimes. For inter-tidal environments in the Windscale (UK) area, Hetherington [27] concluded that the sedimentation of contaminated material was the primary mechanism by which plutonium was incorporated into sediments. In contrast, Livingston and Bowen [35] suggested that upward translocation of plutonium occurred in sediments from the eastern United States, and suggested that bioturbation may play an important role in this process. These latter sediments have a much lower rate of accumulation than the sediments investigated by Hetherington [27], in which bioturbation may well be of little consequence. Aston and Stanners [36] have confirmed and extended Hetherington's observations, and have found no vertical plutonium migration even in sediment cores which exhibit active bioturbation. The question of vertical migration of transuranics is worthy of considerable research in the future.

It may be concluded, therefore, that either singly, or in combination, these two processes (i.e. elemental post-depositional migration and burrowing by organisms) can considerably modify down-core elemental profiles in sediments. It may also be concluded that for the transuranics there is considerable scope for studies on possible vertical migration processes in a wide range of sedimentary environments, and that these studies should extend to elements other than plutonium.

In the proceeding section attention was confined to the analysis of the total non-residual elements, and it was suggested that a general separation of these from the residual elements can yield a great deal of useful environmental data, particularly for the stable trace metals. A simple technique, such as a cold acid leaching, for carrying out this separation is therefore often sufficient for the initial investigation of non-residual trace metal pollution in coastal sediments. However, once such a survey has identified the presence of any 'anthropogenic trace metal fingerprints', it may be necessary to acquire a more detailed knowledge of the location of the metals within the non-residual sediment fraction. Such knowledge is often useful in assessing the potential environmental fate of both trace metals and transuranics.

There is a great variety of chemical techniques which have been designed to establish the distribution of non-residual elements among the components of a sediment — for review, see Chester [37]. The complexity of these techniques depends to a large extent on the nature of the information sought, and it is convenient to relate them to a classification of the non-residual components themselves. Classification schemes for the components of sediments have been proposed by a number of authors (see e.g. Refs [4—7]), and some of these have been used as a basis for chemical separation techniques. For example, Chester and Hughes [6] outlined a technique for the sequential separation of the components of deep-sea sediments which was related to Goldberg's classification. More detailed schemes which have been postulated for the classification of the chemical phases in sediments include those of Gibbs [38] and Förstner and Patchineelam [8]. According to Gibbs [38], the associations of heavy metals with aquatic solids may be divided into four broad types: adsorptive bonding, coprecipitation by hydrous iron and manganese oxides, complexation by organic molecules, and incorporation into crystalline solids. The first three of these are associated with the non-residual sediment fractions and the latter with the residual fraction. Förstner and Patchineelam [8] expanded these categories to include other types of metal bondings which occur in both natural and polluted water systems, and a general scheme based on their classification is given in Table III.

The classification of the trace metal associations in sediments which is listed in Table III forms a useful framework within which it is possible to delineate the types of elements that should be individually separated from coastal and estuarine deposits. Attempts which have been made to chemically separate elements in associations similar to those given in the table include the partition techniques described by Presley et al. [39], Nissenbaum [40], Gibbs [38], Nissenbaum [41], Engler et al. [42], Gupta and Chen [43], and Brannon et al. [44]. If a scheme for the investigation of elemental partitioning in sediments is based on the classification given in Table III, then ideally each reagent used should be specific to the release of elements in one of the various associations. In practice, this is not easy to achieve, and indeed may well be impossible. One reason is that many reagents will affect, at least to some extent, elements bound to the sediment in association with more than one component. However, this can often be overcome by making the chemical separation scheme sequential, i.e. the sediment is progressively leached in such a manner that some components are removed before a particular reagent is used. In this way specific reagents can be made to leach specific components.

TABLE III.   HOST COMPONENTS AND ELEMENTAL BONDING IN
SEDIMENTS[a]

Sediment fraction	Host component	General type of bonding mechanism
Residual	Silicate minerals, heavy minerals	Within mineral lattice structures; e.g. element-oxygen bonds
Non-residual	Hydroxides and oxides of Fe and Mn	Physico-sorption Chemical sorption Coprecipitation
	Calcium carbonate	Physico-sorption Pseudomorphosis Coprecipitation
	Organic matter; e.g. humics, lipids, residual organics	Physico-sorption Chemical sorption Complexation
	Heavy metal precipitates; e.g. hydroxides, carbonates, sulphides	Precipitation

[a] Modified from Föstner and Patchineelam [8].

A technique recently developed at Liverpool [9] will serve as an example of
a sequential scheme designed for the investigation of elemental partitioning in
coastal sediments. The classification of elemental bondings given in Table III was
modified and expressed in a form which was directly applicable to sediment
partitioning studies. The modified classification is listed in Table IV, together
with the reagents used to release elements held in each of the various associations.
Sequentially, the scheme proceeds as illustrated in the flow chart in Fig. 3, and
its application to the distribution of trace metals to a 'polluted' sediment from the
River Mersey (UK) is described below.

The partitioning of a number of trace metals among the various components
of a sample of the Mersey mud is given in Table V, and several conclusions can be
drawn from this data: (a) In this 'polluted' sediment $\gtrsim$ 75% of the total manganese,
zinc, copper, lead and nickel is associated with the non-residual fraction;

TABLE IV. CHEMICAL REAGENTS FOR THE RELEASE OF ELEMENTS FROM HOST COMPONENTS IN SEDIMENTS

Sediment fraction	Elemental association	Leaching agent
Residual	Lattice-held	$HF + HNO_3$
Non-residual	Elements in weakly adsorbed or cation exchangeable forms	$MgCl_2$
	Elements associated with humics etc.	$Na_4P_2O_7$
	Elements associated with carbonates and Fe and Mn hydroxides and oxides	$NH_2OH.HCl + CH_3COOH$
	Elements associated with sulphides and residual organics	H_2O_2

TABLE V. TRACE METAL PARTITIONING IN A MUD FROM THE MERSEY RIVER (UK)

Element	% non-residual	% of non-residual fraction associated with the various sediment phases			
		Weakly exchangeable metals	Humic phases	Carbonate and oxide phases	Sulphide and residual organic phases
Fe	36	0.5	26	38.5	35
Mn	79	4	3	93	trace
Zn	79	trace	12	72	16
Cu	85	6	30	43	21
Pb	80	trace	7	60	33
Ni	91	trace	42	56	trace

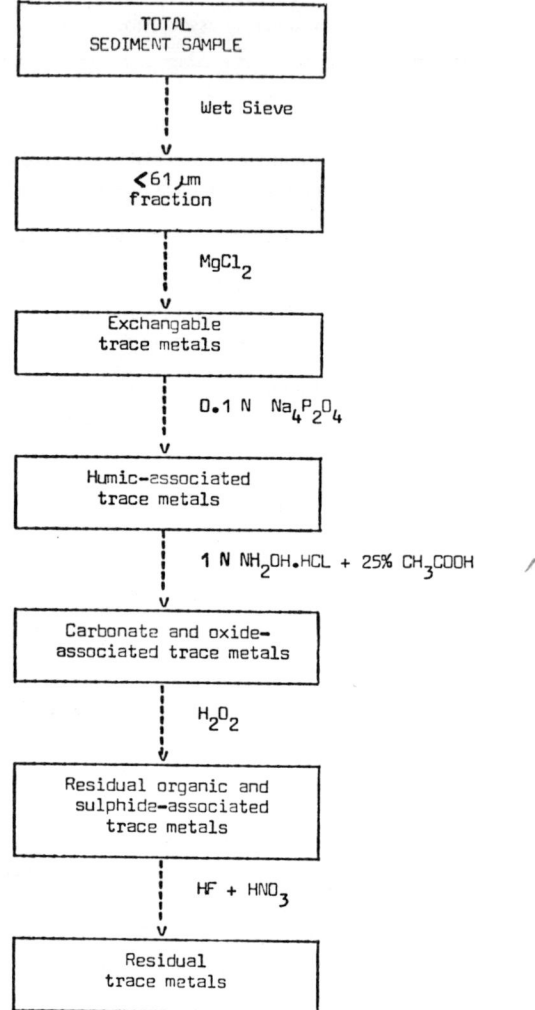

FIG.3. Flow chart for the sequential leaching scheme.

(b) within the non-residual fraction the partitioning characteristics differ from one element to another; (c) the largest fractions of the non-residual manganese, zinc, lead and nickel are located in the component which consists of oxides and carbonates; (d) the organic-associated iron, zinc and copper is partitioned between the humic and the residual organic + sulphide components; (e) the organic-associated nickel is exclusively held in the humic components; (f) the organic-associated lead is strongly concentrated in the residual organic + sulphide component; (g) only trace amounts of the metals studied are in weakly exchangeable positions within the sediment.

Investigations into the chemical partitioning of transuranics in sediments by leaching techniques have been rather limited to date, and have centred on plutonium. For example, Hetherington [28] has shown that plutonium appears to be distributed fairly uniformly throughout all the constituents of the deposited sediments on both the Irish sea-bed and in adjacent estuaries; i.e. the plutonium is not associated with any particular phase, either organic or inorganic. In contrast, the evidence from the very rapid uptake of plutonium to sediments suggests that the plutonium should perhaps be present in association with an inorganic component, e.g. a Fe/Mn hydroxide phase as is found for lake sediments (see e.g. Refs [18, 31, 45]). However, only about 25% of the plutonium in the sediments studied by Hetherington [28] could be leached using the established procedure for the dissolution of iron and manganese described by Chester and Hughes [6]. These observations do not agree with those made recently by Aston and Stanners [36] who studied the association of plutonium with non-detrital Fe/Mn phases in intertidal sediments from the eastern Irish Sea. These authors found that ~ 70 to 100%, with an average of 86%, of the plutonium was in the non-detrital Fe/Mn phase. These recent results are more consistent with the data on plutonium phase associations in lake sediments (see e.g. Refs [18, 31, 45]). Clearly, there is much further work to be undertaken on the partitioning of plutonium and other transuranics in sediments from various environments.

Partitioning studies such as those discussed above are important for the assessment of the subsequent fate of both trace metals and transuranics in the environment. In this context, the organic-metal associations may be of special relevance in determining the extent to which an element becomes available to bottom-feeding biota and so enters the aquatic food chain. This is one area of research which merits more detailed investigation in the future.

3. CONCLUSIONS

Sediments are an important host for trace metals and transuranics in the aquatic environment. However, the residence of the elements in these deposits may only be temporary, and their release into the ecosystem represents a potentially very dangerous hazard. The extent to which this release occurs depends largely on the manner in which the elements are bound to the sediments and on how particular kinds of bonds react to various physico-chemical conditions. It is therefore desirable to have some knowledge of the way(s) in which the elements are bound to the solid components in particular sediment populations.

Chemical separation techniques offer the most useful means for the aquisition of data on the partitioning of trace metals and transuranics in sediments. The complexity of these techniques varies greatly, as does the time and the cost of carrying them out. For some trace metal investigations it is often sufficient to use a very general technique, such as one that merely separates the non-residual from the residual sediment fraction. In others, particularly those involving the transuranics, it is necessary to use more sophisticated techniques for the identification of specific elemental associations within the non-residual fraction itself. A number of chemical partitioning techniques have been described in the preceding sections; however, it is probable that no single technique is 'ideal' for all trace metal and transuranics studies. In practice, the investigator must select the technique best suited to his needs, i.e. to the kind of data he requires, and one of the principal advantages of a sequential leaching scheme, such as that described by Chester et al. [9], is that additional reagents can be 'plugged' in at the appropriate stage in order to gain information on elemental associations which were not included in the initial scheme.

REFERENCES

[1]   OSTERBERG, C., CAREY, A.G., HERBERT, C., Nature **200** (1963) 1270.
[2]   HARVEY, G.R., STEINHAUER, W.G., in Environmental Biogeochemistry, (NRIAGU, J.O. Ed.), Ann Arbor Science Pub., Ann Arbor, Michigan (1974).
[3]   HONJU, S., J. Mar. Res. **36** (1978) 493.
[4]   KRYNINE, P.D., J. Geol. **56** (1948) 130.
[5]   GOLDBERG, E.D., J. Geol. **62** (1954) 249.
[6]   CHESTER, R., HUGHES, M.J., Chem. Geol. **2** (1967) 249.
[7]   ELDERFIELD, H., in Chemical Oceanography, Vol. 5, 2nd Edn (RILEY, J.P., CHESTER, R., Eds), Academic Press, London (1976) 137.
[8]   FÖRSTNER, U., PATCHINEELAM, S.R., Chem. Ztg. **100** (1976) 49.
[9]   CHESTER, R., TOWNER, J., SAYDAM, C., (1980) (in preparation).
[10] BRUTY, D., CHESTER, R., ASTON, S.R., Nature **245** (1973) 73.
[11] PICKERING, R.J., CARRIGAN, P.H., TAMURA, T., AKEE, H.H., BEVERAGE, J.W., ANDREW, R.W., in Disposal of Radioactive Wastes into Seas, Oceans and Surface Waters (Proc. Symp. Vienna, 1966), IAEA, Vienna (1966) 80.
[12] DUURSMA, E.K., BOSCH, C.J., Neth. J. Sea Res. **4** (1970) 395.
[13] TAMURA, T., JACOBS, D.G., Health Phys. **2** (1960) 391.
[14] DUURSMA, E.K., EIGMA, D., Neth. J. Sea Res. **6** (1973) 265.
[15] ASTON, S.R., DUURSMA, E.K., Neth. J. Sea Res. **6** (1973) 255.

[16] PATEL, B., PETEL, S., PAIVAR, S., Estuar. Coast. Mar. Sci. **7** (1978) 49.

[17] JENNE, E.A., WAHLBERG, J.S., USGS Prof. Paper 433-F (1968) pp.16.

[18] EDGINGTON, D.N., ROBBINS, J.A., in Impacts of Nuclear Releases into the Aquatic Environment (Proc. Symp. Otaniemi, 1975), IAEA, Vienna (1975) 245.

[19] SIMPSON, H.J., OLSEN, C.R., TRIER, R.M., WILLIAMS, S.C., Science **194** (1976) 179.

[20] CERRAI, E., MEZZADRI, M.G., TRIULZI, C., Energ. Nucl. **16** (1969) 378.

[21] PILLAI, D.C., DEY, N.N., MATHEW, E., KOTHARI, B.U., in Impacts of Nuclear Releases into the Aquatic Environment (Proc. Symp. Otaniemi, 1975), IAEA, Vienna (1975) 277.

[22] FÖRSTNER, U., WITTMANN, G.T.W., (1979) Metal Pollution in the Aquatic Environment, Springer-Verlag, Berlin (1979) pp. 486.

[23] AGEMAIN, H., CHAU, A.S.Y., The Analyst **101** (1976) 761.

[24] LANDERGREN, S., Rep. Swed. Deep-Sea Exped. **10** (1964).

[25] BOSTRÖM, K., FISHER, D.E., Geochim. Cosmochim. Acta **33** (1969) 743.

[26] HETHERINGTON, J.A., JEFFERIES, D.F., LOVETT, M.B., in Impacts of Nuclear Releases into the Aquatic Environment (Proc. Symp. Otaniemi, 1975), IAEA, Vienna (1975) 139.

[27] HETHERINGTON, J.A., in Environmental Toxicity of Aquatic Radionuclides in the Environment, Ann Arbor Press, Ann Arbor, Michigan (1976) 81.

[28] HETHERINGTON, J.A., Mar. Sci. Commun. **4** (1978) 239.

[29] NOSKIN, V.E., Health Phys. **22** (1972) 537.

[30] BOWEN, V.T., LIVINGSTON, H.D., BURKE, J.C., in Transuranium Nuclides in the Environment (Proc. Symp. San Francisco, 1975), IAEA, Vienna (1976) 107.

[31] ALBERTS, J.J., MULLER, R.N., J. Environ. Qual. **8** (1979) 20.

[32] DUINKER, J.C., van ECK, G.T.M., NOTLING, R.F., Neth. J. Sea Res. **8** (1974) 214.

[33] CHESTER, R., VOUTSINOU, F., (1980) (in preparation).

[34] PRICE, N.B., in Chemical Oceanography, Vol. 6, 2nd Edn (RILEY, J.P., CHESTER, R. Eds), Academic Press, London (1976) 1.

[35] LIVINGSTON, H.D., BOWEN, V.F., Earth Planet. Sci. Lett. **43** (1979) 29.

[36] ASTON, S.R., STANNERS, D.S., Nature (1980) (in press).

[37] CHESTER, R., The Partitioning of Trace Elements in Sediments, Comitato Nazionale Energia Nucleare (1978) pp. 28.

[38] GIBBS, R., Science **180** (1973) 71.

[39] PRESLEY, B.J., KOLODNY, Y., NISSENBAUM, A., KAPLAN, I.R., Geochim. Cosmochim. Acta **36** (1972) 1073.

[40] NISSENBAUM, A., Isr. J. Earth.-Sci. **21** (1972) 143.

[41] NISSENBAUM, A., Isr. J. Earth.-Sci. **23** (1974) 111.

[42] ENGLER, R.M., BRANNON, J.H., ROSE, J., 168th Meeting Am. Chem. Soc., Atlantic City, N.Y. (1974) 17.

[43] GUPTA, S.K., CHEN, K. Y., Environ. Lett.**10** (1975) 129.

[44] BRANNON, J.M., ENGLER, R.M., ROSE, J.R., HUNT, P.G., SMITH, I., in Proc. Conf. Dredging and its Environmental Effects, Amer. Soc. Civil Eng., New York (1976) 455.

[45] EDGINGTON, D.N., ALBERTS, J.J., WAHLGREN, M.A., KARTUNNEN, J.D., REEVE, C.A., in Transuranium Nuclides in the Environment (Proc. Symp. San Francisco, 1975), IAEA, Vienna (1976) 493.

STUDY OF CHEMICAL SPECIATION OF PLUTONIUM IN SEA WATER-SEDIMENT SYSTEM*

E. MATHEW, K.C. PILLAI
Environmental Studies Section,
Health Physics Division,
Bhabha Atomic Research Centre,
Bombay,
India

Abstract

STUDY OF CHEMICAL SPECIATION OF PLUTONIUM IN SEA WATER-SEDIMENT SYSTEM.
Various laboratory studies of simulated Bombay Harbour Bay water and sediments are
discussed with a view to better understanding the chemical speciation of plutonium in the
natural system. Extractions of organically bound plutonium in the sediments with alkali showed
that all the soluble plutonium was converted to an anionic complex form. Results of these
experiments relate the extent of solubilization of plutonium in sea water with varying bicarbonate
concentrations and show that non-dialysable plutonium that is associated with high molecular
weight organic compounds is present in sea water.

1. INTRODUCTION

Transuranic nuclides are present in trace levels in the environment as a result
of nuclear detonations, accidental releases and effluent discharges from irradiated
fuel reprocessing and plutonium handling facilities. Oceans covering 70% of the
earth's surface become a major recipient of these nuclides. Sediments have a
very high accumulation capacity for plutonium and more than 99% of the released
plutonium gets locked up in coastal sediments [1]. The behaviour of low levels
of plutonium remaining in the water is likely to vary, depending upon the environmental conditions and the source of plutonium introduced. The organic
constituents of sea water and sediments are known to solubilize many trace
metals [2] and to influence the uptake of many trace elements by sediments [3].
Sea water, especially coastal waters, is known to contain dissolved humic substances
and organometallic substances [4].

Plutonium forms strong complexes with many organic and inorganic ligands.
Andelman and Rozzell [5] and Polzer [6] have discussed the nature of plutonium
in water environments. Interaction of plutonium with complexing substances in

---

* Part of this work was carried out under IAEA Research Contract No. 1954/RB/RI.

TABLE I. INTERACTION OF PLUTONIUM WITH ORGANIC MATTER (OM)

Experiment	Time after mixing (d)	Insoluble Pu (dis/min per 5 ml)	Soluble plutonium				
			Pu concentration		Non-cationic (%)	Non-anionic (%)	Pu solubilized (μg/ml OM)
			(dis/min per 5 ml)	(M)			
Pure Pu solution + 25 ml organic matter extract (1 ml = 0.37 mg OM)	7	20	68.8	4.3×10^{-10}	100	0	2.7×10^{-4}
pH = 7.0	48	10.6	60.3	3.7×10^{-10}	100	0	2.4×10^{-4}

PILLAI, K.C., MATHEW, E., in Transuranium Nuclides in the Environment, IAEA, Vienna (1976) 25.

soils and natural waters was studied by Bondietti et al. [7]. Association of plutonium with the solubilized soil humic materials was also reported by them. Plutonium is reported [8] to be present in anionic forms in Michigan waters and the major oxidation state of plutonium is the Pu(IV) form. Edgington et al. [9] reported that about 7% of the total sedimentary plutonium was associated with humic and fulvic acids in this lake. However, the increase of organic-bound plutonium with time is not known.

In view of the low levels of plutonium in the waters of the Bombay Harbour Bay, it was not possible to study the nature of dissolved plutonium present in it. Hence, laboratory experiments were carried out to find the species that are formed as a result of interaction of plutonium with organic matter, sea water and sea water containing excess organic matter.

2. INTERACTION STUDIES OF PLUTONIUM

2.1. Interaction with organic matter

Organic matter is extracted from the sediment with alkali mixture ($NaOH + Na_2CO_3$) and purified. The details of the extraction and purification are described elsewhere [10]. Plutonium nitrate solution ($10^{-11}M$) was evaporated to dryness and was allowed to interact with 25 ml of purified organic matter extract (1 ml = 0.37 mg of organic matter). The pH value was maintained at 7.0. Aliquots of the solutions were filtered and dialysed in a 4.8 nm dialyser tubing and the solution was again filtered. All the filtrations in the present study were carried out using 0.22 nm Millipore filter paper. The speciation of plutonium was investigated using cation Dowex-50 (NH_4^+ form) and anion Dowex-1 (OH^- form) ion exchange resins. The plutonium content in the influents and the effluents was estimated by composing the organic matter and measuring the residue. The results are given in Table I. It was found that 1 mg of organic matter solubilized 2.7×10^{-4} μg of plutonium. All plutonium had been converted to an anionic complex form.

2.2. Interaction with sea water

Interaction of plutonium with sea water was studied using plutonium solutions as well as low-level liquid effluents containing plutonium from fuel reprocessing operations. Plutonium solution was evaporated and the residue was taken in a few drops of 0.1N HNO_3 and mixed with filtered sea water. Aliquots of the samples were removed at known intervals and filtered. Plutonium in the filtrate was estimated after the addition of ^{236}Pu tracer.

TABLE II. SOLUBILITY AND SPECIATION OF PLUTONIUM IN SEA WATER AT DIFFERENT INTERVALS OF TIME

II.A. Pure Pu solution + Millipore filtered sea water, pH 7.4–7.8 throughout the experiment

Time after mixing (d)	Insoluble Pu (dis/min per 100 ml)	Insoluble (%)	Soluble plutonium				
			Pu concentration (dis/min per 100 ml)	(M)	Non-cationic (%)	Non-anionic (%)	Non-ionic (%)
0	186.1	55.5	149.4	4.6×10^{-11}	–	46	9
1	245.8	65.5	129.5	4.0×10^{-11}	32	26	15
10	261.0	81.1	60.6	1.9×10^{-11}	62	60	42
30	264.4	86.1	42.6	1.3×10^{-11}	61	64	63
120	338.6	92.1	32.1	1.0×10^{-11}	60	82	–

II.B. Pure Pu solution + Millipore filtered sea water + 20 mg organic matter, mixed pH 7.4–7.8 throughout the experiment

Time after mixing (d)	Insoluble Pu (dis/min per 100 ml)	Insoluble (%)	Pu concentration (dis/min per 100 ml)	(M)	Non-cationic (%)	Non-anionic (%)	Non-ionic (%)
0	88.2	27.8	228.4	7.1×10^{-11}	38	24	15
1	81.0	20.6	311.2	9.7×10^{-11}	45	42	23
10	83.0	17.6	389.5	12.1×10^{-11}	71	38	28
30	108.6	30.4	248.1	7.7×10^{-11}	83	49	34
150	110.3	27.0	304.0	9.5×10^{-11}	79	53	28

The species formed was studied by passing part of the filtrate through cation Dowex-50 (Na^+) and anion Dowex-1 (Cl^-) columns. The effluents were analysed separately. The non-ionic (neutral) species were estimated in the effluents obtained from cation + anion columns.

2.3. Interaction with sea water containing organic matter

The above experiment was repeated using sea water to which about 20 mg (per litre of sea water) of organic matter were added. The results obtained for interactions with sea water and sea water containing excess organic matter are given in Table II. A maximum amount of plutonium solubilized was obtained soon after mixing, however there was a decrease with time. At pH 7.4 to 7.8, the soluble plutonium concentration in sea water was $10^{-11}M$. Both cationic and anionic species decreased with time. In the case of sea water to which organic matter was added, more plutonium was solubilized with time and remained in solution. It was observed that after 30 days there was a tendency for plutonium to precipitate out, probably along with organic matter which was observed to be precipitated on the filter bed. The results after 150 days, however, indicate further solubilization of plutonium, possibly from that adsorbed on the glass surface. There was an increase in the anionic species with time which was due to organic matter added to sea water which was about two orders of magnitude higher than the natural levels present in sea water. This was required since the amount of plutonium added to sea water in the laboratory experiments was also comparatively higher.

In the case of effluents, known aliquots were mixed with different volumes of filtered sea water and the experiments were carried out similarly. The results obtained are given in Table III. The maximum solubilization of plutonium observed was about 43 dis/min per litre ($10^{-12}M$), which was an order of magnitude lower than when pure plutonium solution interacted with sea water. The lower concentration of plutonium is the result of the removal of plutonium along with other fission products present in the effluent. Therefore, the introduction of such effluents into the aquatic environment is likely to result in lower concentrations of plutonium in soluble form in sea water. The species obtained were more or less the same as in the case of pure plutonium solutions. The non-ionic species were less compared with those obtained at higher plutonium concentrations. The ageing effect was also less in this case. In the case of effluents that had interacted with sea water containing organic matter it was observed that cationic species were not detectable and about 90% were in an anionic complex form.

In all these experiments plutonium sorption on glass surface has been observed. The calculations were made on the basis of plutonium in solution.

TABLE III. SOLUBILITY AND SPECIATION OF PLUTONIUM (CONTAINED IN EFFLUENTS) IN SEA WATER AT DIFFERENT INTERVALS OF TIME

| Experiment | Time after mixing | Soluble plutonium | | Non-cationic (%) | Non-anionic (%) | Non-ionic (%) |
		Pu concentration (dis/min per 100 ml)	(M)			
1 ml effluent + 200 ml filtered sea water	15 min	4.4	1.4×10^{-12}	55	59	14
1 ml effluent + 200 ml filtered sea water	15 d	4.3	1.3×10^{-12}			
1 ml effluent + 200 ml filtered sea water	15 months	4.1	1.3×10^{-12}	88	63	50
1 ml effluent + 200 ml filtered sea water + 3.5 mg organic matter	15 months	4.8	1.5×10^{-12}	100	17	17
5 ml effluent + 100 ml filtered sea water	1 d	1.8	5.6×10^{-13}			
10 ml effluent + 200 ml filtered sea water	3 d	1.8	5.6×10^{-13}	63	63	26
1 ml effluent + 1000 ml filtered sea water	7 d	1.0	3.1×10^{-13}	86	45	31

PILLAI, K.C., MATHEW, E., in Transuranium Nuclides in the Environment, IAEA, Vienna (1976) 25.

TABLE IV. INFLUENCE OF CARBONATE, BICARBONATE AND ORGANIC
MATTER ON SOLUBILIZATION OF Pu IN FILTERED SEA WATER

Experiment	Soluble plutonium		
	(dis/min per 100 ml)	(M)	(μg Pu/ltr)
Sea water with low $CO_3^=$, (3.6 mg/ltr) pH: 7.45	60	1.9×10^{-11}	4.4×10^{-3}
Sea water with low $CO_3^=$, HCO_3^- + 20 mg organic matter	180	5.6×10^{-11}	13.1×10^{-3}
Sea water with normal $CO_3^=$, HCO_3^- concentration (24.6 mg/ltr) pH: 7.9	72	2.2×10^{-11}	5.3×10^{-3}
Sea water with normal $CO_3^=$, HCO_3^- concentration + 20 mg organic matter	212	6.6×10^{-11}	15.5×10^{-3}
Sea water with excess $CO_3^=$, HCO_3^- (96.36 mg/ltr) pH: 8.2	141	4.4×10^{-11}	10.3×10^{-3}
Sea water with excess $CO_3^=$, HCO_3^- + 20 mg organic matter	512	15.9×10^{-11}	37.3×10^{-3}

3. SOLUBILIZATION STUDIES OF PLUTONIUM IN SEA WATER WITH VARIOUS CARBONATE, BICARBONATE CONCENTRATIONS AND ORGANIC MATTER

Andelman and Rozzell [5] reported that the addition of carbonate significantly increases the solubilization of Pu(IV) and (VI) in sea water owing to the formation of carbonate complexes. However, the effect of organic matter in the presence of carbonate was not studied.

Experiments were carried out to study the extent of solubilization of plutonium in sea water with varying carbonate, bicarbonate contents. Sea water samples (250 ml) containing various carbonate, bicarbonate concentrations — low (by acidification and boiling), normal and excess (adding 18 mg $NaHCO_3$) sea water content — were prepared and mixed with the same amount of plutonium solution. After a day's equilibration, two 100 ml aliquots were made from each

TABLE V. NON-DIALYSABLE PLUTONIUM IN SEA WATER

| Experiment | dis/min per 100 ml | | | % |
	Soluble	Non-dialysable	Insoluble Pu after dialysis	Pu Non-dialysable
Natural sea water with excess carbonate and organic matter added	256	199.2	9.7	77.8
Natural sea water with excess carbonate added	66.6	9.3	30.4	13.1
Organic matter removed from sea water	22.3	Nil	Nil	Nil

solution and to one 1 ml (2 mg) of organic matter and to the other 1 ml of distilled
water was added and again kept for a day with intermittent shaking.

The soluble plutonium present in each case in the filtrate was estimated and
the results obtained at various carbonate, bicarbonate concentrations are shown in
Table IV. The plutonium solubilized is more or less directly proportional to the
carbonate added. The amount of plutonium solubilized was 5.7×10^{-5} $\mu g/mg$
of carbonate added. The solubilization of plutonium is much higher if organic
matter is added to sea water. The amount of plutonium solubilized per mg of
organic matter added was 5×10^{-4} $\mu g/ltr$ of sea water. However, in the presence
of excess carbonate, bicarbonate the solubilization of plutonium appeared to
be higher. At a carbonate, bicarbonate concentration of 96 mg/ltr ($10^{-3} M$),
plutonium was solubilized to the extent of 13×10^{-4} $\mu g/mg$ of organic matter
added to sea water.

A known volume of the above filtrate with and without organic matter was
dialysed against distilled water. The solution from the dialysing tube was filtered
and the plutonium in the filtrate was estimated. Results are shown in Table V.
A larger amount of plutonium remained in a non-dialysable fraction when
organic matter was added to the sea water. The natural sea water showed a
relatively small amount of non-dialysable plutonium owing to the small amounts
of organic matter being naturally present in it. This was confirmed by dialysing
two aliquots of sea water freed from organic matter (by adding 50 mg aluminium
nitrate solution per litre sea water and filtering it). Following interaction with
plutonium solution the filtrate was dialysed. In this case the non-dialysed sea water
did not indicate the presence of any soluble plutonium.

This clearly indicates that non-dialysable plutonium in association with high
molecular weight organic compounds can be present in sea water. This is significant
especially in coastal waters, where higher concentrations of organic matter in the
form of humic materials are present in sea water and the potential for contamination
is high.

4. ORGANICALLY BOUND PLUTONIUM AND AMERICIUM IN COASTAL SEDIMENTS

Organic matter was extracted from a sediment collected from Trombay Naval
Jetty, a location near to the discharge point of the Research Centre. The extract
was purified by repeated dialysis and ion-exchange. The plutonium and americium
content in the purified organic matter was then estimated and the results are
shown in Table VI. The presence of both plutonium and americium in the
purified organic matter was evident from the result. Organically bound plutonium
was also found to increase with time. The association of plutonium with naturally
dissolved organic substances in sea water as well as the presence of bound

TABLE VI. PLUTONIUM IN ORGANIC FRACTION OF COASTAL SEDIMENTS

Year of sampling	Sediment for organic extraction (g)	Total in sediment (dis/min)		Organic matter extracted (mg)	Pu in purified organic matter extract		Am in purified organic matter extract	
		Pu	Am		(dis/min)	(dis/min per mg)	(dis/min)	(dis/min per mg)
1968	200	20.0	-	1080	0.27	2.46×10^{-4}	-	-
1970	33	45.2	-	-	0.18	-	-	-
1975	20	18.3	-	196	0.31	1.59×10^{-3}	-	-
1976	1000	-	-	919	2.20	2.39×10^{-3}	-	-
1977	1090	1014	578	4283	43.3	1.02×10^{-2}	1.71	4.01×10^{-4}

Sampling Location: Trombay Naval Jetty.

TABLE VII.   TRACER RECOVERIES OBTAINED FOR TWO OXIDATION STATES

Sample No.	Sample volume (ltr)	^{242}Pu(IV) fraction		^{236}Pu(VI) fraction	
		% recovery	% ^{236}Pu(VI) tracer present	% recovery	% ^{242}Pu(IV) tracer present
1	1	68.6	1	70.1	4.3
2	1	61.2	1.2	66.6	6.5

plutonium with the organic matter fraction of the contaminated sediment may result in the biological cycling in the marine environment.

5.   VALENCY OF PLUTONIUM AND RELEASE FROM SEDIMENTS

Some preliminary studies on the valency states and release of plutonium from sediments were made. Valency states of plutonium in sea water were determined as two fractions, either (IV) or (VI), with the (IV) fraction made up of both (III) and (IV) valence states, while the (VI) fraction consisted of both (V) and (VI) valence states.

Sea water was kept in contact with a plutonium contaminated sediment for a period of one year in a tub. During this period the sediment had undergone redox changes, evident from the occasional release of H_2S. The overlying sea water was syphoned out and filtered through Whatman 42. The method followed for separation and estimation of different valency states of plutonium was that of Lovett and Nelson [11]. ^{242}Pu(IV) and ^{236}Pu(VI) tracers were added simultaneously to sea water. Pu(IV) was precipitated on NdF_3, keeping Pu(VI) in solution by the oxidant potassium dichromate solution. Pu(VI) was later reduced to (IV) and then precipitated on NdF_3. Plutonium in two fractions was separated by ion exchange, and measured by alpha spectrometry.

Tracer recoveries obtained for two valency states are shown in Table VII. Percentage recoveries obtained for ^{242}Pu(IV) and ^{236}Pu(VI) were about 65 and 68%, respectively. The amount of ^{239}Pu present in sea water in two valency states is shown in Table VIII. Both valency states of ^{239}Pu were found in the sea water samples in the same proportion. The total amount of plutonium in sea water (0.15−0.17 dis/min per litre) was higher than one year earlier (0.11 dis/min per litre) indicating a release of plutonium from the sediment. Further work is in progress to gain better understanding of the release mechanism.

TABLE VIII. ^{239}Pu PRESENT IN SEA WATER IN TWO VALENCY STATES

Sample No.	^{239}Pu in (IV) valency (dis/min per ltr)	^{239}Pu in (VI) valency (dis/min per ltr)
1	0.076 ± 0.02	0.073 ± 0.02
2	0.07 ± 0.02	0.1 ± 0.02

The interaction of plutonium with organic matter and bicarbonates in sea water shows increased solubilization and complexation and changes in the speciation of plutonium in sea water. Introduction of effluents containing trace levels of plutonium will further alter these interactions depending upon the nature of plutonium in effluents and its chemical constituents. The accumulation of plutonium in the organic fraction of the sediment with time and the possibilities of release of plutonium from sediments demonstrate the need for further studies on the bio-geochemistry of plutonium.

ACKNOWLEDGEMENTS

We acknowledge the valuable guidance given by Dr. A.K. Ganguly, National Fellow in Environmental Sciences, Department of Science and Technology, and Shri S.D. Soman, Head, Health Physics Division. The assistance obtained from Shri Kuttappan in analysis and Shri N.N. Dey while preparing the paper is also gratefully acknowledged.

REFERENCES

[1] PILLAI, K.C., MATHEW, E., GANGULY, A.K., Studies on Pu in the Marine Environment, Rep. BARC/I-462 (1977).
[2] KOSHY, E., DESAI, M.V.M., GANGULY, A.K., Studies on organo-metallic interactions in the marine environment part-I. Interactions of some metallic ions with dissolved organic substances in sea water, Curr. Sci. 38 (1969).
[3] GANAPATHY, S., PILLAI, K.C., GANGULY, A.K., Adsorption of Trace Elements by Near Shore Sea Bed Sediments, Rep. BARC-376 (1968).
[4] DESAI, M.V.M., GANGULY, A.K., "Organic and organo-metallic substances in the marine environment", presented at the Symp. on the Interaction Between Water and Living Matter, Odessa, 1975.
[5] ANDELMAN, J.B., ROZZELL, T.C., "Plutonium in the water environment", Radio-nuclides in the Environment 8, Advances in Chemistry Ser. 93 (1970) 118.

[6] POLZER, W.L., Solubility of Plutonium in Soil/Water Environment, Rep. HASL-83401 (1971).

[7] BONDIETTI, E.A., REYNOLDS, S.A., SHANKS, M.H., "Interaction of plutonium with complexing substances in soils and natural waters", Transuranium Nuclides in the Environment (Proc. Symp. San Francisco, 1975), IAEA, Vienna (1976) 273.

[8] WAHLGREN, M.A., ALBERTS, J.J., NELSON, D.M., ORLANDINI, K.A., "Study of the behaviour of transuranics and possible chemical homologues in Lake Michigan water and biota", Transuranium Nuclides in the Environment (Proc. Symp. San Francisco, 1975), IAEA, Vienna (1976) 9.

[9] EDGINGTON, D.N., ALBERTS, J.J., WAHLGREN, M.A., KARTTUNEN, J.O., REEVE, C.A., "Plutonium and americium in Lake Michigan sediments", Transuranium Nuclides in the Environment (Proc. Symp. San Francisco, 1975), IAEA, Vienna (1976) 493.

[10] KOSHY, E., GANGULY, A.K., Organic Materials in the Marine Environment and their Interactions with some Metal Ions, Rep. BARC 402 (1968).

[11] LOVETT, M.B., NELSON, D.M., "The determination of the oxidation states of plutonium in sea water and associated particulate matter", presented at the Symposium on the Determination of Radionuclides in Environmental and Biological Materials, CEGB, London (1978).

OBSERVATIONS ON THE DEPOSITION, MOBILITY AND CHEMICAL ASSOCIATIONS OF PLUTONIUM IN INTERTIDAL SEDIMENTS

S.R. ASTON, D.A. STANNERS
Department of Environmental Sciences,
University of Lancaster, Lancaster,
United Kingdom

Abstract

OBSERVATIONS ON THE DEPOSITION, MOBILITY AND CHEMICAL ASSOCIATIONS OF PLUTONIUM IN INTERTIDAL SEDIMENTS.

The deposition of plutonium in intertidal sediments is considered together with the possible post-depositional migration and chemical associations of plutonium. While there is good agreement that plutonium is rapidly and effectively removed from sea water to sediments in the vicinity of nuclear effluent discharges to sea, there is some contrasting evidence on its post-depositional behaviour in intertidal and coastal sediments. Recent results from the Ravenglass estuary adjacent to the Windscale reprocessing facility have shown a lack of plutonium mobility in intertidal sediments. These studies have also allowed a value for the *apparent* diffusion coefficient of plutonium in sediments to be derived, and preliminary observations on the association of plutonium with non-detrital Fe/Mn phases.

1. INTRODUCTION

Several aspects of the behaviour of plutonium in sediments are important in the evaluation of the fate of this transuranic in the natural environment. Here, some aspects of the behaviour of plutonium in marine and intertidal sediments are considered together with a preliminary evaluation of the geochemical association of this element in the intertidal sedimentary environments. The literature on the marine sedimentary geochemistry of plutonium is not extensive, and the following discussions are limited to the small number of observations on plutonium in near-shore and estuarine (intertidal) deposits.

2. PLUTONIUM DEPOSITION AND MIGRATION IN MARINE SEDIMENTS

The three main sources of information on the deposition and migration of plutonium in intertidal sediments polluted by plutonium from the nuclear fuel cycle are Hetherington [1, 2], and Aston and Stanners [3]. These studies relate to sediments collected in the vicinity of the Windscale nuclear fuel reprocessing

TABLE I. THE ANNUAL DISCHARGES OF PLUTONIUM (^{238}Pu AND $^{239+240}$Pu) TO THE IRISH SEA FROM WINDSCALE DURING 1968–1978 (Ci·a^{-1})

Year	Discharge	Year	Discharge
1968	828	1974	1248
1969	816	1975	1195
1970	936	1976	1272
1971	1128	1977	984
1972	1548	1978	1672
1973	1776		

(Data from Hetherington et al. [4] and BNFL [5].

plant, Cumbria, England. The sediment cores in the three studies were collected in the Ravenglass estuary, a small inlet about 10 km to the south of the Windscale facility.

Hetherington [1] reported that in each of the core samples collected in 1973 the absolute activities of $^{239+240}$Pu and the ratio of $^{239+240}$Pu/^{238}Pu activities changed with depth, decreasing and increasing respectively with increasing depth and an approximately exponential fashion. This pattern of plutonium deposition was attributed to the increasing discharges of plutonium from Windscale to the Irish Sea, and to the modifications to the type of plutonium being reprocessed (Table I). Hetherington concluded that the sedimentation of contaminated material was the primary mechanism by which plutonium was incorporated into the deposited sediments. The rapid removal of plutonium to sedimentary material from sea water has been demonstrated [1, 6]. Hetherington's data which were based on cores collected in 1973, with an increasing trend in plutonium activities towards the sediment surface, did not prove that the plutonium reflected the changing discharges from Windscale or a lack of post-depositional mobility. The 1973 core profiles may have been produced by the upward translocation of plutonium by migration and/or mixing (*apparent diffusion processes*). Other workers [7, 8] have attributed the plutonium profiles in sediments from the near-shore, continental shelf and continental slope areas of the eastern United States of America to upward translocation by biological sediment mixing and chemical mobilization processes. These authors have pointed out that the importance of biological mixing processes to the upward movement of plutonium in sediments may be less for faster accumulating deposits, e.g. estuarine sediments.

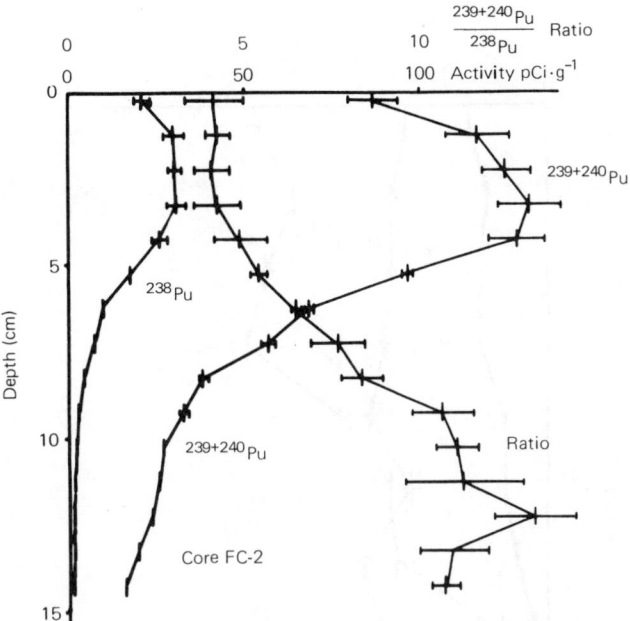

FIG.1. Depth distributions of $^{239+240}Pu$, ^{238}Pu activities and the $^{239+240}Pu/^{238}Pu$ ratio in sediment core FC-2, Newbiggin, Ravenglass Estuary. (All errors not shown in this and subsequent figures are <5%.)

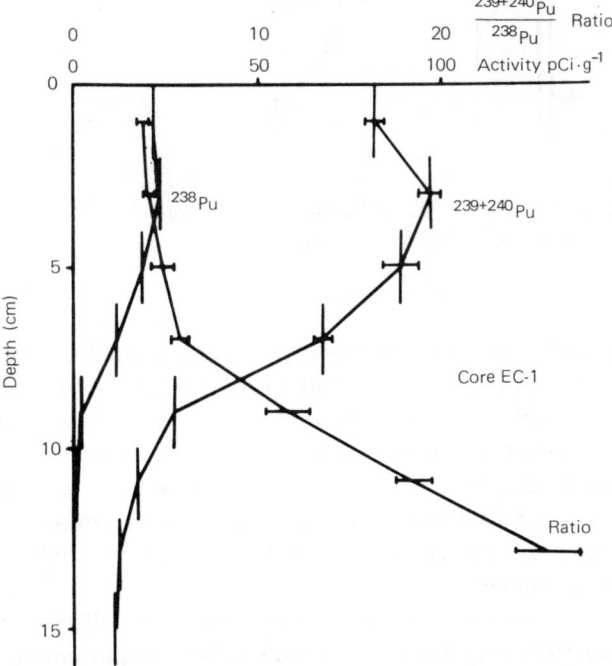

FIG.2. Depth distributions of $^{239+240}Pu$, ^{238}Pu activities and the $^{239+240}Pu/^{238}Pu$ ratio in sediment core EC-1, Newbiggin, Ravenglass Estuary.

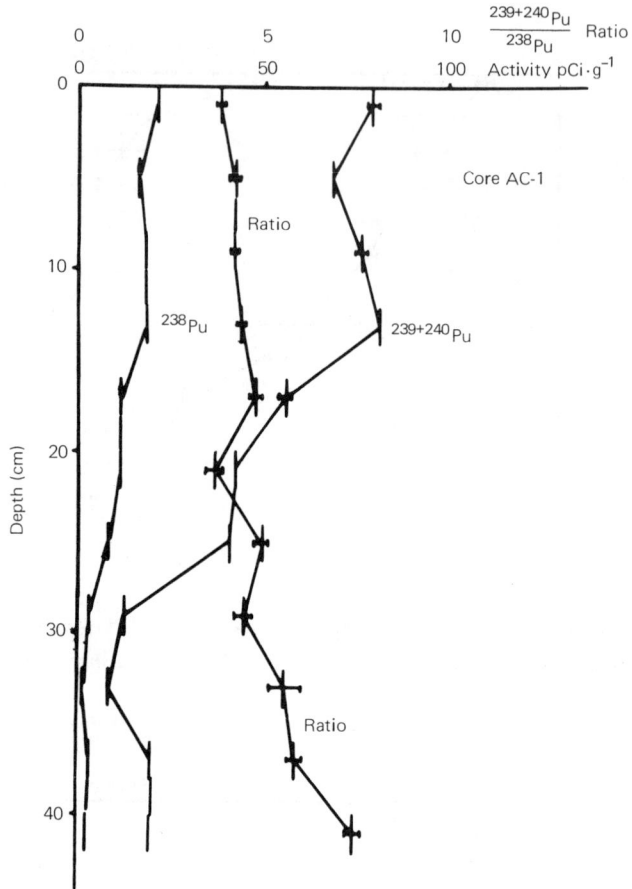

FIG.3.   *Depth distributions of* $^{239+240}Pu$, ^{238}Pu *activities and the* $^{239+240}Pu/^{238}Pu$ *ratio in sediment core AC-1, Newbiggin, Ravenglass Estuary.*

Recent studies by Aston and Stanners [3] have extended the work of Hetherington on the deposition and mobility of Windscale plutonium in the Ravenglass estuary. The sediment cores used by these authors were collected in 1978, and thus reflect the incorporation of plutonium in sediments before, during and after the peak discharge in 1973. This allows a more substantial appraisal of Hetherington's hypothesis which was based on samples which coincided with the peak discharges. The results of the recent studies are discussed in some detail below.

Figures 1, 2 and 3 show plutonium profiles for cores collected from Ravenglass during 1978 [3]. The profiles show maximum plutonium activities at depth in the vertical sediment profiles. Exponential increases in plutonium

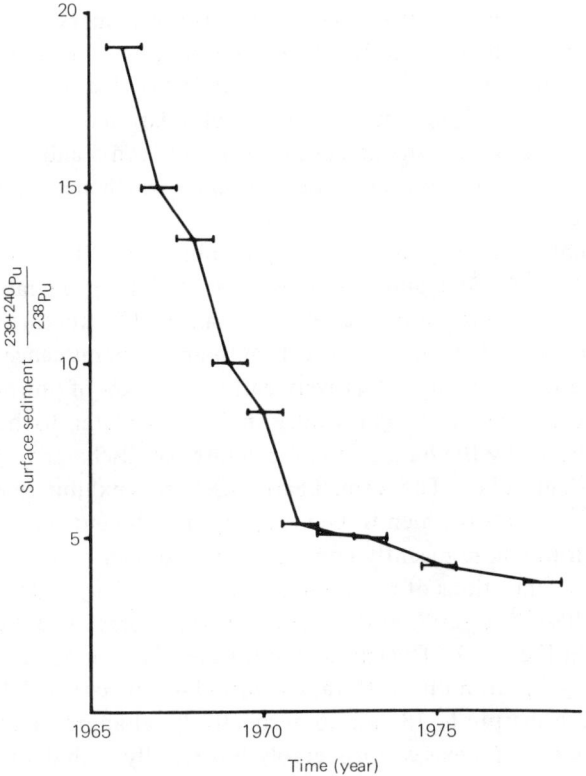

FIG.4.   The average $^{239+240}Pu/^{238}Pu$ activity ratio in surface sediments from Newbiggin, Ravenglass Estuary, as a function of time.

content towards the surface of the sediments are not observed, the maximum activities occur at 3, 3 and 13 cm in cores FC-2, EC-1 and AC-1, respectively. This contrasting pattern of plutonium deposition is easily accounted for by the changes in the plutonium discharges from Windscale (Table I). This represents a considerable decrease in plutonium discharges since the peak value of discharge in 1973 which coincided with Hetherington's sampling period. Aston and Stanners' [3] core profiles clearly show that the continuing temporal pattern of plutonium discharge is preserved within the intertidal sediment profiles. The presence of the 'buried' plutonium peak activities within the cores confirms Hetherington's suggestion that sedimentation of contaminated material is the primary mechanism.

Evidence of the presence, or absence, of the post-depositional migration of plutonium in intertidal sediments by such processes as bioturbation, diffusion, advection, etc. may be obtained from the examination of the isotopic ratio

$^{239+240}$Pu/^{238}Pu at different depths. This ratio has varied markedly with time
for the plutonium discharge from Windscale, and indirect, but useful, evidence
for these changes is shown by the $^{239+240}$Pu/^{238}Pu ratio of surface sediments
(<1 cm) collected over a long time period. A compilation of such data is
given in Fig.4. For example, the higher ratios of between 5 and 20 which
occurred in the 1960s have now decreased to values which are relatively
constant at about 4.

Environmental processes are exceedingly unlikely to differentiate the
isotopes $^{239+240}$Pu and ^{238}Pu, and this allows the $^{239+240}$Pu/^{238}Pu ratio to be used to
specify the age of the plutonium in a sediment sample. Furthermore, an
important application of the ratio is in determining the significance of vertical
sediment mixing within a core. Relatively early discharges of plutonium from
Windscale, e.g. pre 1968, were significantly smaller than later discharges.
This is clearly shown by the data on the mean rates of discharge to the Irish Sea
given by Hetherington [1]. These small early discharges exhibit characteristic
high $^{239+240}$Pu/^{238}Pu ratios which to be preserved in sediments must not have
been mixed with the more recently contaminated sediments which contain
much greater concentrations of plutonium with substantially lower ratios.

The $^{239+240}$Pu/^{238}Pu ratios at different depths in cores FC-2, EC-1, AC-1
are also shown in Figs 1–3. The surface ratio values in each of the cores represent
recently discharged plutonium, with ratios around 4. In cores FC-2 and EC-1
the ratios increase sharply to 13 and 26, respectively, at approximately 13 cm.
In core AC-1 the ratio increases considerably less rapidly with depth, reaching a
value of only 7.5 at 41 cm. The preservation of the older high ratios in
sediments contaminated with small amounts of plutonium shows that mixing
with the sediments containing much greater amounts of low $^{239+240}$Pu/^{238}Pu
ratio plutonium is negligible. This is a clear indication that the vertical migration
of plutonium or the downward mixing of sediment grains contaminated with
plutonium are not important. This conclusion is consistent with the observed
profiles of plutonium activities in the cores which reflect the temporal variations
in discharge. Aston and Stanners [3] reported an *apparent* diffusion coefficient
of <10^{-12} cm$^2 \cdot$s^{-1} for plutonium in these sediments. This value was calculated
from the method of Aston and Stanners [9].

3.   CHEMICAL FORMS AND EXTRACTION OF PLUTONIUM

The deposition of plutonium in the intertidal sediments is clearly related
to the sedimentation of contaminated particles. In the northeastern Irish Sea,
the deposition of plutonium from sea water is rapid and effective; e.g.
Hetherington [1] has estimated that about 95% of discharged plutonium is
removed to sediments in the immediate vicinity of Windscale. Also, the

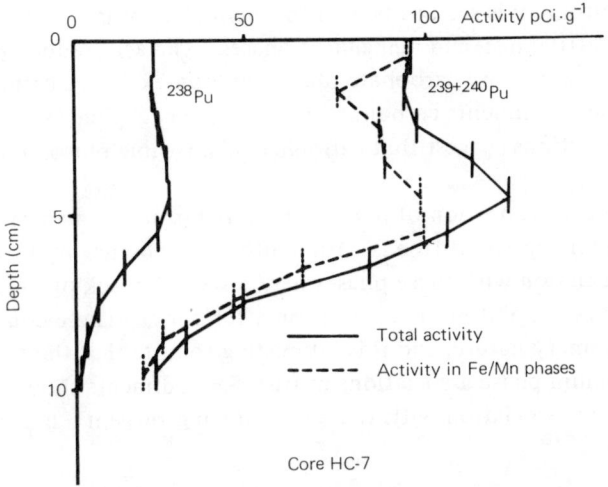

FIG.5. The depth distributions of $^{239+240}Pu$ and ^{238}Pu activities, and fraction of $^{239+240}Pu$ activity in non-detrital Fe/Mn phases, in sediment core HC-7 Newbiggin, Ravenglass Estuary.

oxidation state of plutonium in Irish Sea suspended solids has been determined [10]. The Pu(IV) state was found to be dominant in particulates, while Pu(VI) was abundant in the dissolved plutonium. Aston [11] has used theoretical arguments to evaluate the chemical forms of plutonium in sea water, and has shown that the Pu(IV) state will be favoured as the highly insoluble [Pu(OH)$_4$]. There is good evidence that Pu(OH)$_4$ is strongly adsorbed and coprecipitated onto negatively charged surfaces [12], and the iron and manganese oxide/hydroxide phases of marine particulates should provide excellent scavenging for the [Pu(OH)$_4$]. Indeed, coprecipitation of plutonium by ferric hydroxide has been used for the analytical preconcentration of this element from large samples of sea water (see e.g. Ref. [13]).

As a preliminary test of the association of plutonium with non-detrital iron and manganese oxide/hydroxide phases, the extraction technique of Chester and Hughes [14] was applied by Aston and Stanners [3]. The acid/ reducing agent mixture of acetic acid and hydroxylamine hydrochloride was found to remove virtually all of the oxide/hydroxide phase, and associated trace metals. The extraction method was applied to a core (HC-7) from Newbiggin, Ravenglass Estuary, and the results are shown in Fig.5. The percentages of $^{239+240}Pu$ and ^{238}Pu removed by the acid/reducing reagent ranged over 78 to 97% and 74 to 100%, respectively. The mean percentages of $^{239+240}Pu$ and ^{238}Pu removed were 84.7 and 84.6%. Thus, a large proportion of the total plutonium is seen to be in a form which appears to be associated with the non-detrital iron and manganese phases. The acid/reducing reagent technique will also remove carbonate phases and ion-exchange cations. However, in these sediments carbonate minerals are negligible ($<2\%$) and theoretical predictions suggest that cationic exchangeable plutonium will not exist [11].

Although the proportion of plutonium in the iron and manganese phases varies, there are no apparent trends with depth. The absence of a trend in plutonium association with these phases may reflect the lack of significant diagenetic changes in plutonium association after burial. These conclusions are of a very preliminary nature, and it is interesting to note that the only other study of plutonium phase associations in Irish Sea sediments showed only ~25% plutonium association with the acid/reducing reagent removable phase [2].

REFERENCES

[1] HETHERINGTON, J.A., "The behaviour of plutonium nuclides in the Irish Sea", Environmental Toxicity of Aquatic Radionuclides: Models and Mechanisms (MILLER, M.W., STANNARD, J.N., Eds), Ann Arbor Science, Michigan (1976).

[2] HETHERINGTON, J.A., The uptake of plutonium nuclides by marine sediments, Mar. Sci. Commun. 4 (1978) 239.

[3] ASTON, S.R., STANNERS, D.A., The deposition and immobility of plutonium in intertidal sediments from the Irish Sea, Nature, Lond. (submitted) (1980).

[4] HETHERINGTON, J.A., JEFFERIES, D.F., MITCHELL, N.T., PENTREATH, R.J., WOODHEAD, D.S., "Environmental and public health consequences of the controlled disposal of transuranic elements to the marine environment", Transuranium Nuclides in the Marine Environment (Proc. Symp. San Francisco, 1975), IAEA, Vienna (1976) 139.

[5] BNFL, Annual Report on Radioactive Discharges and Monitoring of the Environment, British Nuclear Fuels Ltd., Risley, England (1978, 1979).

[6] LIVINGSTON, H.D., BOWEN, V.T., Windscale effluents in the waters and sediments of the Minch, Nature, Lond. 269 (1977) 586.

[7] LIVINGSTON, H.D., BOWEN, V.T., "Americium in the marine environment — relationships to plutonium", Environmental Toxicity of Aquatic Radionuclides: Models and Mechanisms (MILLER, M.W., STANNARD, J.N., Eds), Ann Arbor Science, Michigan (1976).

[8] LIVINGSTON, H.D., BOWEN, V.T., Pu and ^{137}Cs in coastal sediments, Earth Planet. Sci. Lett. 43 (1979) 29.

[9] ASTON, S.R., STANNERS, D.A., The determination of estuarine sedimentation rates by $^{134/137}$Cs and other artificial radionuclide profiles, Estuar. Coastal Mar. Sci. 9 (1979) 529.

[10] NELSON, D.M., LOVETT, M.B., Oxidation state of plutonium in the Irish Sea, Nature 276 (1978) 599.

[11] ASTON, S.R., Evaluation of the chemical forms of plutonium in sea water, Mar. Chem. 8 (1980) 319.

[12] KRAUS, K.A., "Hydrolitic behaviour of the heavy elements", Peaceful Uses of Atomic Energy (Proc. Int. Conf. Geneva, 1955) 7, UN, New York (1956) 245.

[13] MURRAY, C.N., FUKAI, R., Measurement of $^{239+240}$Pu in Northwestern Mediterranean, Estuar. Coastal Mar. Sci. 6 (1978) 145.

[14] CHESTER, R., HUGHES, M.J., A chemical technique for the separation of ferro-manganese minerals, carbonate minerals and adsorbed trace elements from pelagic sediments, Chem. Geol. 2 (1976) 249.

TECHNIQUES AND METHODS OF INTERFACIAL TRANSFER AND TRANSPORT PROCESSES IN THE SEDIMENT-WATER LAYER

E.K. DUURSMA
Delta Institute for Hydrobiological Research,
Yerseke,
Netherlands

Abstract

TECHNIQUES AND METHODS OF INTERFACIAL TRANSFER AND TRANSPORT PROCESSES IN THE SEDIMENT-WATER LAYER.
The fate of radionuclides in aquatic systems is linked with sediments and particulate matter, as accumulation factors of 10^4 to 10^5 are noted. It is questioned whether the methodologies used at present to determine the short-term behaviour of radionuclides are also applicable for long time periods. These techniques are evaluated and recommendations are made to broaden their utility for the understanding of both short and long-term behaviour.

1. INTRODUCTION

The fate of radionuclides in aquatic systems is to a great extent connected with particulate matter and sediments [1]. Many radionuclides can be accumulated to a factor of 10^4 to 10^5 calculated for the same volume of sediment and water.

Thus, a 20 cm bottom-sediment layer is a potential sink for a large water column from which radionuclides are ultimately precipitated. For accumulation factors of 10^4 to 10^5 this water column is theoretically 2000 to 20 000 m high for a stagnant system, while for currents flowing over a sediment a similar picture can be drawn. For example, the same 20 cm sediment layer would be capable of accumulating radionuclides from a 2-m-thick current with a length of 1000 to 10 000 m.

The first example may apply to the ocean, where only some tenths of centimetres of sediment are sufficient to accumulate all radionuclides from the ocean water. The second example shows that, for example in the rivers or estuaries with strong currents, the sediment top layer may be quickly saturated and any further uptake by the sediment from the currents is controlled by the penetration into the bottom. This may be relatively rapid if sediments are stirred up so that scavenging can occur but, on the other hand, very slow in undisturbed sediment layers.

These examples demonstrate clearly that immobilization of radionuclides and redistribution over different matrixes are 'four'-dimensional processes related to space and time.

At present our knowledge is based on only thirty years of experience since fallout and liquid waste disposal studies started. The problems we are still facing are those of long-term reactions, burial and redistribution of radionuclides for periods of centuries and thousands of years. The question is whether the technical approach of the determination of radionuclide behaviour during short time periods (decades) is basically applicable to these long-term periods.

A short evaluation of the techniques involved in such studies of interfacial transfer and intersedimentary transport processes is given.

2.   TECHNICAL PROBLEMS TO DETERMINE TRANSFER AND TRANSPORT PROCESSES

2.1. Water to sediments

2.1.1. Diffusion of dissolved substances

Transfer of only dissolved material from a water body to the bottom sediments is a process which is still little understood. In complete stagnant water without any convection, molecular-diffusion models can be used to calculate the amount of dissolved material available to be absorbed at the water-sediment interface. Problems arise when this diffusion is faster than the sediment interface can accumulate and/or redistribute deeper into the bottom [2].

This problem is even larger for non-stagnant waters in which convection and eddy diffusion are the dominant transport processes. Although this generates a much higher chance of contact between water and bottom, the transfer through the water-sediment interface is probably again the limiting factor.

The techniques available to determine the eddy diffusion processes in the water phase are those used in physical oceanography and limnology. If salinity gradients occur, these parameters can be used, while otherwise studies have to be carried out with radioactive or non-radioactive tracers. An example is the application of tritium spikes at different depths to determine vertical diffusion [3] in a whole lake. In fact, any dissolved natural or artificial substance would be suitable if a gradient or imposed gradient can occur in the water column.

The main problem which remains is to translate the results from one dissolved substance to another, since reactions at the water-sediment interface are selective.

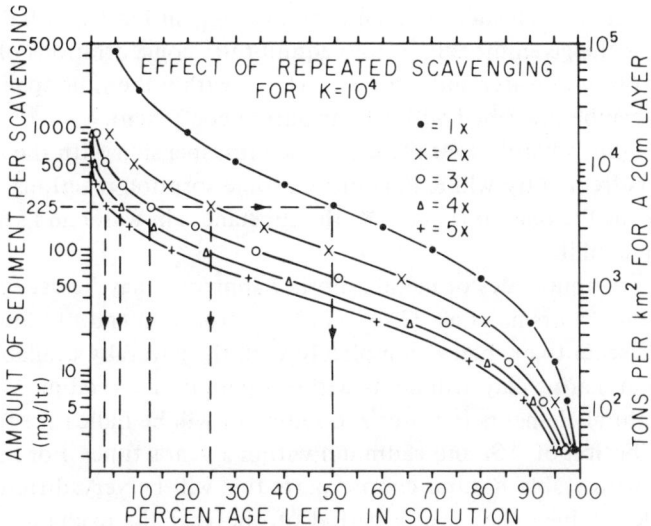

FIG.1. *The effect of five successive scavengings by equal amounts of sediment with a density (dry) of 2.27 g · ml⁻¹ for a radionuclide with a distribution coefficient of 10⁴.*

2.1.2. Precipitation of suspended substances

Geosecs studies [4], measurements by Duinker and Nolting [5] and experiments by Aston and Duursma [6] have demonstrated that, in addition to heavy metals and radionuclides attached to or incorporated in particulate matter, a micro-particulate fraction occurs of possibly hydroxides or hydrocarbonate-metal (radionuclide) compounds.

The vertical transport to the bottom may occur through different pathways such as settling with particles, scavenging by settling particles and precipitation with faecal pellets after passage through living organisms [7]. The velocity of settling differs greatly depending on specific weight, grain size and agglomeration of the particles and turbulence of the water system. Faecal pellets can settle at rates of between 50 and 300 m · d⁻¹ [8, 9], while scavenging with resuspended sedimentary particles may follow a pattern as is given in Fig.1 [10].

The techniques available for the study of these problems are for settling material the sediment-trap techniques [11] as used by the Woods Hole Group of Honjo, Farrington and co-workers, and the conventional filtration and centrifugation techniques. The trap technique is only applicable for ocean systems, since in estuaries and coastal areas the turbulence due to currents is an interfering factor.

The model approach such as given for scavenging in Fig.1 is still difficult to effectuate since the given model assumes equilibrium concentrations of the substances involved in water and sediment which, within the time span of the resuspension, can be described with a distribution coefficient.

In nature this is hardly ever the case, since resuspension with the same material occurs frequently while, in addition, some sorption reactions are time dependent. Thus the reactions between the substances in water and sediment may not reach equilibrium.

For some radionuclides or metals another approach may be used like, for example, the one Duursma et al. [12] found for strontium and cadmium. Both these elements seem to exchange completely with the naturally available strontium and cadmium in marine clay sediments within a period of several months. Thus the ratio of natural elements in water and sediment will be indicative for the ultimate distribution of ^{90}Sr and cadmium within a year's time. For most other radionuclides with stable natural counterparts this will be very different, since it was determined that over a period of one year the exchange was incomplete to only a small fraction. The exchange is either extremely slow, or limited to the surface of sedimentary particles. In the latter case there are different types of reactions due to the differences in speciation of the contaminant in relation to the natural counterpart.

3.   TRANSFER THROUGH THE INTERFACIAL LAYER

3.1. Physico-chemical diffusion

There are hardly any techniques available to study experimentally diffusion in sediment other than to simulate cores and very thin layers of sediment upon which a gradient of concentration is imposed. It is difficult to define an interfacial layer, since in quiescent conditions an interface is only imaginary. Diffusion can occur into or out of the sediment from or to the water, respectively [13]. Subsequently the measurements etc. are equal as will be discussed in section 4.1. Under conditions of turbulence, the interface may be defined as that layer which is being reworked. This will be treated in section 3.2. below.

3.2. Reworking and biological transport

A major transfer of substances through the water-sediment interface occurs by either physical or biological reworking processes of the top layer of the sediment. Currents, waves and organisms can temporarily stir up a few centimetres of sediment which become suspended in the supernatant water. Additionally, benthic organisms living in burrows sheltered in the bottom feed on sedimentary material or plankton from or above the interface. Although in many cases faeces are redeposited on the interface, part of the substances may find their way by this route into the bottom sediments.

The reworking of the top-layer of the sediment can be determined by tracer experiments using radionuclide [14] and dye-attached sediments, as well as by measuring the activities of organisms in situ and in experiments [15, 16]. All these processes are generally short-term ones (months, years) and can therefore be studied either in situ or with model experiments. The principal basis for most techniques is that a process with natural or artificial spikes can be followed from a determined moment (t_0) so that successive movements of these spikes can be followed.

Whatever technique or application of spikes is used, or subsampling is performed, it is most essential that the complete experiment be performed according to a mathematical model. Only then factors like apparent diffusion coefficients can be determined. Experiments thus performed in situ with coloured natural sediment [17] showed that a natural population of zoobenthos could achieve a reworking with diffusion coefficients of 10^{-8} to 10^{-7} cm$^2 \cdot$ s^{-1}. This apparent diffusion is only valid for the first period of the experiment, since later 'diffused' particles are returning to original positions.

Another approach to be used is the determination of the thickness of the reworked sediment layer in cm $\cdot$ a^{-1} as a mixing parameter. If the reworked depth is known, this result might also be expressed in a turnover time. Such results can be obtained from measurements on, for example, benthic worms from which the individual reworking activity can be determined [16].

The integrated effect can be studied in some cases from field situations. Such a possibility is presented by the Irish Sea, where the effluent radionuclides from the Windscale plant can be followed. For ^{106}Ru Duursma and Gross [10] evaluated the vertical diffusion with a coefficient of 10^{-7} cm$^2 \cdot$ s^{-1}, which is at least 10^3 times too high to be explained by physico-chemical diffusion only. Thus, reworking of the sediment must have occurred either by currents and waves or by organisms.

The techniques applied are those of analysis of cores and comparison of the results obtained as they can be reconstructed on the basis of sorption and diffusion experiments [18].

4. TRANSPORT IN THE BOTTOM LAYER

4.1. Physico-chemical diffusion

In quiescent sediment layers transport can only occur via the interstitial water passing through the pores. For those sediments it is the physico-chemical diffusion process that, in the absence of any biological activity other than that of bacteria, determines the transport.

The techniques for the determination of such diffusion processes can apply spikes of radionuclides to experimental cores or study the gradients of substances as determined in natural cores. However, for transport with very low diffusion coefficients the experiments are too time-consuming and the processes have to be studied by indirect methods on the basis of sorption-disorption equilibria between substances in the interstitial water and sedimentary particles [18].

Although for a number of radionuclides results were obtained for various marine sediments [19], the diffusion coefficients thus obtained have only significance for periods of at most some decades. This could be demonstrated, for example, with ^{65}Zn which nuclide continued to be absorbed within the crystal lattices of clay [12, 20]. The result is that diffusion coefficients may be lower than determined when the processes of absorption by sedimentary particles continue for decades or centuries.

Although this demonstrates that the methodology is incomplete, it still has the advantage of providing information on the kind of dominating transfer process in sediments. As already mentioned in section 3.2 for ruthenium, physical reworking and bioturbation processes can thus be distinguished, since knowledge exists as to what physico-chemical diffusion would be able to produce (see Ref.[21]).

4.2. Bioturbation

In natural aquatic sediments, bioturbation processes of two types can be expected: (i) bioturbation of the interstitial water and (ii) bioturbation of the sediment particles. The latter, as far as sediment reworking is concerned, has already been treated in section 3.2.

The bioturbation of interstitial water has as yet attracted little attention, although more than 10^6 meiofauna organisms per m^2 (harpacticides, nematodes) can be present in the first 20 cm of a sediment layer [22]. Such bioturbation should enforce the random motion in interstitial water and cause higher diffusion rates. This indeed was detected by Vanderborght and Billen [23] for NO$_3^-$ in bottom sediments. They calculated diffusion coefficients of 10^{-4} cm$^2 \cdot$ s^{-1}, which is a factor of thirty higher than molecular diffusion would permit. Meiofauna may well be responsible for this phenomenon.

In order to prove the effects of meiofauna under different conditions, experiments should be set up in which the activities of the organisms are studied in detail. Possibly a functional relation may be detected between the induced diffusion with the oxygen demand of these organisms and thus the thickness of the oxygenated top layer. For radionuclides this has an additional implication since the speciation may be dependent on the redox system.

Bioturbation with macrozoobenthos has in addition to reworking of sediments another aspect which is that of burrows like tubes, holes and larger pores. These make the sediment more permeable for either dissolved or particulate matter. Organisms can play an active role in this respect by pumping water through these burrows and thus imposing an exchange between surface water and horizons inside the bottom sediments.

The $^{239, 240}$Pu profiles as determined by Livingston and Bowen [24] may thus be explained. Probably, micro-size particles containing these isotopes are thus transported deeper into the sediment. As explained earlier for ruthenium physico-chemical diffusion could not be held responsible since plutonium adsorbs to sediment with sorption-distribution coefficients of 10^4, which results in a diffusion coefficient of less than 3×10^{-10} cm$^2 \cdot$ s^{-1} [25].

Also for these processes the techniques do not apply other methodologies than determinations carried out on cores and experiments with sediments in suspension or thin layers.

5.   GENERAL CONCLUSIONS

(a)   The techniques of studying transport processes in water, water-sediment interface and bottom sediments are either based on (i) classical sampling of these phases and studying the gradients of natural parameters including radio-nuclides or spikes (dyes, radionuclides), or (ii) experiments with minor quantities of sediment to study the water-sediment sorption/desorption processes in addition to selected diffusion processes.

(b)   Generally, the techniques applied to aquatic sediments produce results which are applicable for short time periods (months, years and sometimes a few decades).

(c)   For the evaluation of long-term processes (centuries and longer) only techniques used for studying natural processes are available. Until now the implications of these studies have been little elaborated.

(d)   It is strongly recommended that research is focused on the problems of long-term (i) accumulation of radionuclides in estuaries and coastal areas, and (ii) deposition of low and high-level radioactive wastes in deep-sea sediments.

REFERENCES

[1]   DUURSMA, E.K., SMIES, M., "Sediments and transfer at and in the bottom interfacial layer", Pollutant Transfer and Transport in the Sea (KULLENBERG, G., Ed.), CRC Press Inc., West Palm Beach, USA, 1980 (in press).

[2]   LERMAN, A., Geochemical Processes; Water and Sediment Environments, John Willey & Sons, New York (1979) pp. 481.

[3] QUAY, P.D., BROECKER, W.S., HESSLEIN, R.H., FEE, E.J., SCHINDLER, D.W.,
     "Whole lake tritium spikes to measure horizontal and vertical mixing rates",
     Isotopes in Lake Studies (Proc. Advisory Group Meeting Vienna, 1977), IAEA, Vienna
     (1979) 175.

[4] CHESSELET, R., "Modes of settling and organic input to the sediment seawater inter-
     face", Colloque Int. CNRS Biogéochimie de la matière organique à l'interface eau-
     sédiment marin, extended Abstract (1976).

[5] DUINKER, J.C., NOLTING, R.F., Distribution model for particulate trace metals in the
     Rhine estuary, Southern Bight and Dutch Wadden Sea, Neth. J. Sea Res. 10 (1976) 71.

[6] ASTON, S.R., DUURSMA, E.K., Concentration effects on ^{137}Cs, ^{65}Zn, ^{60}Co and ^{106}Ru
     sorption by marine sediments with geochemical implications, Neth. J. Sea Res.
     6 (1973) 225.

[7] FOWLER, S., "Biological transfer in the watercolumn", Pollutant Transfer and Transport
     in the Sea (KULLENBERG, G., Ed.), CRC Press Inc., West Palm Beach, USA, 1980.

[8] TURNER, J.T., Sinking rates of fecal pellets from the marine copepod *Pontella meadii,*
     Mar. Biol. 40 (1977) 249.

[9] EPPLEY, R.W., PETERSON, B.J., Particulate organic matter flux and planctonic new
     production in the deep ocean, Nature 282 (1979) 677.

[10] DUURSMA, E.K., GROSS, M.G., "Marine sediments and radioactivity", Radioactivity in
     the Marine Enrivonment, Natl. Acad. Sci. Wash. D.C. (1971) 147.

[11] ZEITSCHEL, B., DIEKMANN, P., UHLMANN, L., A new multisample sediment trap,
     Mar. Biol. 45 (1978) 285.

[12] DUURSMA, E.K., DAWSON, R., ROS VICENT, J., Competition and time of sorption
     for various radionuclides and trace metals by marine sediments and diatoms, Thalass.
     Yugosl. 11 (1975) 47.

[13] MATISOFF, G., Time dependent transport in Chesapeake Bay sediments: Part 1.
     Temperature and chloride, Am. J. Sci. 280 (1980) 1.

[14] INTERNATIONAL ATOMIC ENERGY AGENCY, Tracer Techniques in Sediment
     Transport, Technical Reports Series No. 145, IAEA, Vienna (1973) pp. 234.

[15] RHOADS, D.C., Organism-sediment relations on the muddy seafloor, Oceanogr. Mar.
     Biol. Ann. Rev. 12 (1974) 263.

[16] CADÉE, G.C., Sediment reworking by *Arenicola marina* on tidal flats in the Dutch
     Wadden Sea, Neth. J. Sea Res. 10 (1976) 440.

[17] FRANCKE, J.W., SMIES, M., "Bioturbation in sediment cores from a mudflat in the
     Eastern Scheldt", in Progress Report 1979 Delta Inst. Hydrobiol. Res., Verhand. Kon. Akad.
     Wet., afd. Nat. 2e reeks (1980) (in preparation).

[18] DUURSMA, E.K., BOSCH, C.J., Theoretical, experimental and field studies of radio-
     isotopes concerning diffusion in sediments and suspended particles in the sea.
     Part B: Methods and experiments, Neth. J. Sea Res. 4 (1970) 395.

[19] DUURSMA, E.K., EISMA, D., Theoretical, experimental and field studies concerning
     reactions of radioisotopes with sediments and suspended particles of the sea. Part C:
     Applications to field studies, Neth. J. Sea Res. 6 (1973) 265.

[20] ROS VICENT, J., COSTA YANGUE, F., PARSI, P., STATHAM, G., DUURSMA, E.K.,
     The ease of release of some trace metals and radionuclides being sorbed for prolonged
     periods by marine sediments (unpublished).

[21] ASTON, S.R., STANNERS, D.S., The determination of estuarine sedimentation rates
     by ^{134}Cs/^{137}Cs and other artificial radionuclide profiles, Estuar. Coast. Mar. Sci
     9 (1979) 529.

[22] WILLEMS, K., SANDEE, A.J.J., "Meiozoobenthos: density and biomass", in "Progress Report 1978, Delta Inst. Hydrobiol. Res., Vehand. Kon. Ned. Akad. Wet, afd. Nat.2e reeks, **73** (1979) 168.

[23] VANDERBORGHT, J.P., BILLEN, G., Vertical distribution of nitrate concentration in interstitial water of marine sediments with nitrification and denitrification, Limnol. Oceanogr. **20** (1975) 953.

[24] LIVINGSTON, H.D., BOWEN, V.T., Pu and ^{137}Cs in coastal sediments, Earth Planet. Sci. Lett. **43** (1979) 29.

[25] DUURSMA, E.K., PARSI, P., "Distribution of plutonium-237 between sediment and seawater", XXIVe Congrès Assemblée plénière CIESM, Monaco, Rapp. Proc. Verb. des Réunions **23** 7 (1976) 159.

SEDIMENTS AS INDICATORS OF ARTIFICIAL RADIONUCLIDE DISTRIBUTION WEST OF LA HAGUE

P. GUÉGUÉNIAT
CEA, IPSN/SERE,
Laboratoire de Radioécologie Marine,
Département de Protection,
Cherbourg,

J.P. AUFFRET
Laboratoire de Géologie Marine,
Université de Caen,
Caen,

J. BALLADA
CEA, IPSN,
Département de Protection,
Fontenay-aux-Roses,
France

Abstract

SEDIMENTS AS INDICATORS OF ARTIFICIAL RADIONUCLIDE DISTRIBUTION WEST OF LA HAGUE.
   Four-year investigations of the coastal surface sediments and associated radionuclides in the English Channel show complex distribution patterns around La Hague. Plutonium ratios in the particulate fraction confirm the general understanding gathered from long-term gamma studies — mainly that accumulation of La Hague generated radionuclides is in the Bay of Mont-Saint-Michel and along the Normandy-Brittany Coast. The oil from the Amoco Cadiz accident caused a significant increase in coastal sediment activity of [144]Ce. Experimental studies with plutonium and hydrocarbons indicate that plutonium is more difficult to desorb than [144]Ce.

1.   INTRODUCTION

   Artificial radionuclides released in the Irish Sea and the Engish Channel from the La Hague and Windscale nuclear fuel reprocessing plants have been measured in the marine environment. These data provide valuable hydrological information [1—4].
   Studies were conducted between 1976 and early 1979 to determine the distribution of the main gamma-emitting artificial radionuclides in coastal sediments of the English Channel, the North Sea and the Atlantic Ocean. The resulting data were used to describe radioactivity transfers from the La Hague plant in the northwest Cotentin [5] and to assess the radiological impact of the

March 1979 Amoco Cadiz oil spill which occurred in portions of the area considered [6]. The goal was to summarize the knowledge acquired on gamma-emitting radionuclides between 1976 and 1980, and to examine the initial data on plutonium distribution in sediments as a means of investigating radioactivity transport west of La Hague. In addition, this paper presents an experimental study on the plutonium transport in the presence of oil spills such as that caused by the Amoco Cadiz accident, based on investigations of sediments to the extreme west of Brittany.

2.   DISTRIBUTION OF ^{144}Ce AND ^{106}Ru IN ENGLISH CHANNEL SEDIMENTS

Radioactive releases from the La Hague plant between 1974 and 1980 mainly contained ^{106}Ru and, to a lesser extent, ^{144}Ce. Maximum releases of ^{144}Ce (half-life: 290 days) occurred during the winter of 1974–1975, the ^{106}Ru being discharged at more regular intervals ranging from 2000 to 5000 Ci quarterly, except during the last quarter of 1978 (10 600 Ci).

Cerium exists in the coastal waters essentially in the particulate state, whereas ^{106}Ru is a complex mixture of soluble, semi-soluble, colloidal and particulate forms which vary according to the type of wastes considered. This distinction is particularly important because the dispersion of radionuclides from the La Hague plant discharge structure was represented by a preferential transport of soluble elements to the east and deposit of particulates to the west [5] in the Normandy-Brittany Gulf and, more specifically, in the Bay of Mont-Saint-Michel. The western boundary of impact for the releases was established at the Bay of Lannion (Fig. 1). The coastal area further west of this boundary was subject only to fallout from nuclear weapons testing (18th − 22nd Chinese nuclear tests for the period investigated).

Three reference zones were determined to study the main transport characteristics of artificial radionuclides in the English Channel using coastal sediments as indicators: the northwest Cotentin coastal area, the Bay of the Seine and the Bay of Mont-Saint-Michel.

In the northwest Cotentin, measurements were taken at regular intervals in the port of Fermanville, which represents the eastern boundary (50 km from discharge structure) beyond which ^{144}Ce activity drops sharply to a level comparable with that observed in the areas subject only to nuclear fallout. The activity levels at Fermanville do not exceed those observed in other regions of the northwest Cotentin closer to the discharge structure. However, the data were found to follow uniform patterns which can be directly correlated with the releases.

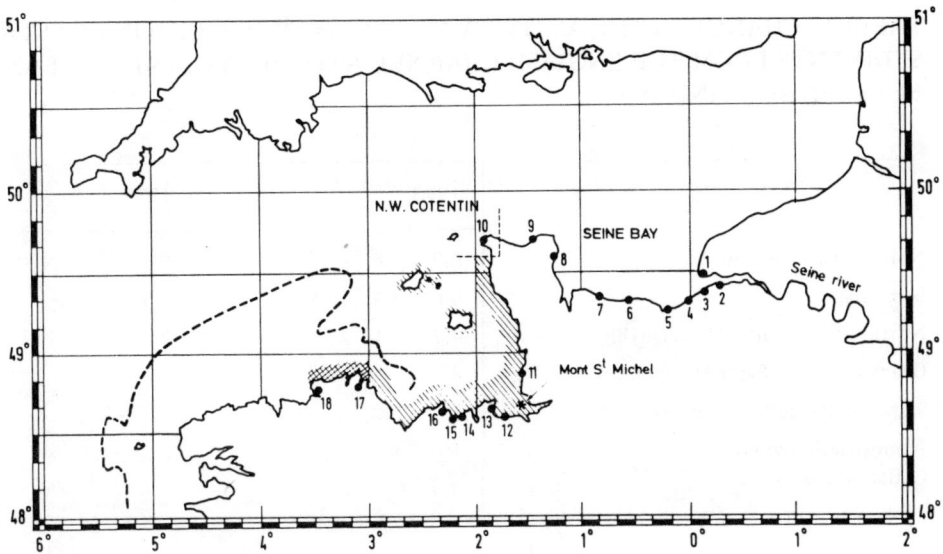

FIG.1.   Location of station:

(1) = Le Havre	(10) = Goury
(2) = Honfleur	(11) = Granville
(3) = Touques	(12) = Le Vivier
(4) = Dives	(13) = Cancale
(5) = Ouistreham	(14) = Fremur
(6) = Courseulles	(15) = Arguenon
(7) = Port-en-Bessin	(16) = Frênaye Bay
(8) = Saint-Vaast	(17) = Trieux
(9) = Férmanville	(18) = Lannion

Impact of industrial wastes on coastal sediments, this aera represents the Normandy-Brittany Gulf.

Western limit of the impact of industrial wastes.

Map of the largest spreading of Amoco Cadiz oil from March 17 to April 28, 1978.

The analysis of marine sediments as indicators described below supports the hypothesis of different dispersion patterns for the soluble and insoluble forms of the radionuclides released.

2.1. Experimental method

The distribution of artificial radionuclides in coastal sediments was determined with particular attention to the estuaries, bays and ports where silt accumulates. Samples taken at each site were classified according to characteristics of the sedimentary environments: total reservoirs, mud flats, salt meadows, and holes of

TABLE I. CHANGE IN THE ACTIVITY OF ^{144}Ce (+ ^{144}Pr) (pCi/g) IN COASTAL
SEDIMENTS FROM THE ENGLISH CHANNEL AND THE ATLANTIC OCEAN
BETWEEN 1977 AND 1979

	1977	1979
North − Pas-de-Calais	0.9 − 1.6	0.5 − 1.2
Seine Bay	0.6 − 1.5	0.8 − 1.3
North-west Cotentin (Fermanville)	4.7 − 9.1	2 − 3
Granville (Mont-Saint-Michel Bay)	8.2	1.1 − 2.1
Total Mont-Saint-Michel Bay	5 − 8.2	1 − 2
Atmospheric fallout (Atlantic coast)	1 − 1.5	0.5 − 1.5

accidentally trapped water. The samples were collected at depths ranging from
surface level to 10 cm and the representativity of this procedure was verified by
examination of vertical radioactivity profiles.

The radioactivity of the samples was analysed with Ge(Li) detectors using
gamma counting techniques at the Atomic Research Unit of the French Navy in
Cherbourg. The ^{144}Ce concentrations reported also take into account ^{144}Pr, a
daughter product with the same energy level; similarly for ^{106}Ru and ^{106}Rh.

2.2. ^{144}Ce distribution

Table I lists the mean activity levels of ^{144}Ce in February 1977 at the three
reference sites, as well as the data collected in the area subject only to nuclear
fallout. The activity levels observed in the Bay of the Seine (0.6−2.2 pCi/g) are
lower than in the Bay of Mont-Saint-Michel (5.5−8.2 pCi/g) and represent an
order of magnitude comparable to fallout levels (1.5 pCi/g upper boundary).
Consequently, ^{144}Ce releases are not detectable in most of the Bay of the Seine
samples collected, despite the 160 Ci discharged from the La Hague plant in 1976,
the 360 Ci discharged in 1978 and the 360 Ci discharged again in 1979. Analysis
of the 1974−1975 data leads to the same conclusions, but the number of samples
was too limited at that time to justify a definitive statement. It should be noted
that the ^{144}Ce properties of the samples collected in the eastern English Channel
and the North Sea coastal regions are the same as for the Bay of the Seine.

The curves in Fig. 2 show the variations in ^{144}Ce activity in the waters near
the discharge point (in pCi/ltr) and in the sediments (mean activity in pCi/g) of
the ports of Granville (Bay of Mont-Saint-Michel) and Fermanville.

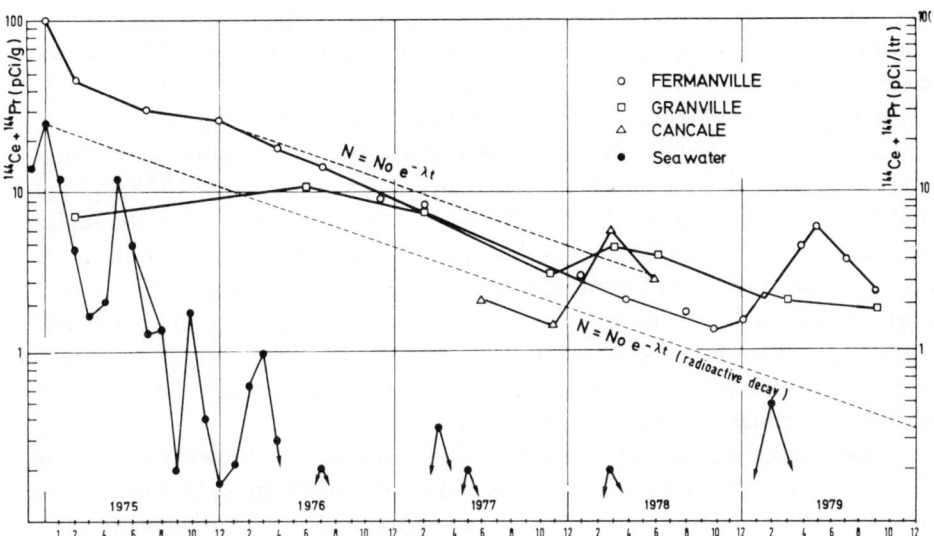

FIG.2. Activities of ^{144}Ce (+ ^{144}Pr) in Cape La Hague sea water (pCi/ltr) and Fermanville, Granville, Cancale sediments (pCi/g).

2.2.1. Northwest Cotentin waters

The chemical separation procedure used was described by Guéguéniat and Gandon [7]. A known quantity of H_2O_2 is added directly to the sea water samples and MnO_2 is precipitated to adsorb ^{144}Ce through reduction of MnO_4K. Prior to coprecipitation, the heaviest suspended solids are removed after settling for three hours. As shown in Fig. 2, the analyses revealed that the activities recorded during the winter of 1974–1975 (up to 30 pCi/ltr) are considerably higher than in the following years (below 1 pCi/ltr), even after adjustment for decay.

2.2.2. Fermanville and Granville port sediments

At Fermanville, the effects of the releases were virtually instantaneous. In January 1975, the ^{144}Ce induced specific activities of up to 100 pCi/g. Immediately thereafter, decay effects were observed in a pattern which became exponential from January 1976 through October 1978 with a half-life of 255 days (99.8% correlation factor), i.e. shorter than for ^{144}Ce (290 days).

At Granville, after peaks at an indeterminate period and difficult to assess due to lack of data between February 1975 and June 1976, a drop from 11 to 3 pCi/g was initially recorded from June 1976 through November 1977, followed by an increase from 3 to 4.6 pCi/g between November 1977 and March 1978, and a drop between February 1977 and March 1980. At Cancale, another regularly monitored site in the Bay of Mont-Saint-Michel, a comparable distribution is observed with an even more pronounced increase in March 1978 (2.2 to 6 pCi/g). Consequently, in March 1978, more ^{144}Ce was deposited in the surface sediments of the Bay of Mont-Saint-Michel in contrast with the diminishing activity in the northwest Cotentin, which was observed starting in 1975. This phenomenon can be explained by the arrival of moving silt — usually in suspended form or deposited in the subtidal zone — leading to the conclusion that the effects of a specific release may be felt in the surface sediments of the Bay of Mont-Saint-Michel's intertidal zone several years later.

Analysis of vertical radioactivity profiles indicates the presence of ^{144}Ce in the Bay of Mont-Saint-Michel to depths of 1.60 m, which shows that large masses of sediment are deposited every year in this region (1 500 000 m^3). Sedimentation at Fermanville and in the Bay of the Seine is, however, considerably lower, the ^{144}Ce being detectable in the first 20 centimetres and the first 10 centimetres, respectively.

2.3. Distribution of ^{106}Ru

Table II shows that distribution of ^{106}Ru in sediments collected at Fermanville is proportional to the releases. In May 1979, a maximum of 91 pCi/g was observed, followed by a relatively fast drop (to 30 pCi/g in January 1980).

Table III compares ^{106}Ru activity in the sediments of the Bay of Mont-Saint-Michel (Granville) and the Bay of the Seine (Saint-Vaast-La-Hougue-Ouistreham) from 1976 to 1980. Surprisingly, the eastern measurement values were lower than in the west from 1976 to 1978, and slightly exceed the latter in 1980. There are two possible explanations for this:

(a) the ^{106}Ru solubilization with time observed in the immediate vicinity of the discharge point (50—90% in particulate form during the winter of 1974—1975, less than 10% in 1979—1980 independent of weather conditions) causes increasingly large movements of the radionuclide to the east;

(b) the effects of the exceptionally high ^{106}Ru releases during the last quarter of 1978, rapidly observed to the east during May 1979, had not been completely felt to the west at the beginning of 1980. This hypothesis is consistent with the potential time lapse of three years between ^{144}Ce release and its effects on the west coast.

TABLE II.   MEAN ACTIVITY (pCi/g) OF ^{106}Ru + ^{106}Rh IN SEDIMENTS
FROM FERMANVILLE

	Releases (Ci/a)	Mean activity
11.12.74		30
27.01.75	1975 − 22 425	11.7
23.07.75		11.6
03.12.75		10.9
15.04.76	1976 − 15 000	9.8
06.07.76		8.4
08.11.76		7
03.02.77	1977 − 15 000	10.4
11.01.78	1978 − 21 700	6.1
27.04.78	of which 10 550	7.6
11.08.78	during the	9.2
16.10.78	last quarter	8.2
15.12.78		13.9
22.04.79	1979 − 20 000	55.0
16.05.79		91.4
26.07.79		47.6
06.09.79		34.8
11.01.80		29.4

   Ongoing studies in 1980–1981 should provide more detailed information on
the influence of the two parameters considered above. With regard to physico-
chemical impact, it should be noted that the various forms of ^{106}Ru in sea water
are dependent on equilibrium reactions [8]. As a consequence, the soluble forms
should yield insoluble forms after disruption of equilibrium caused by the disappearance of insoluble forms readily adsorbable in the marine environment.

TABLE III. MEAN ACTIVITY (pCi/g) OF ^{144}Ce (+ ^{144}Pr), ^{125}Sb, ^{106}Ru (+ ^{106}Rh), ^{137}Cs IN SEDIMENTS FROM ST-VAAST, OUISTREHAM AND GRANVILLE BETWEEN 1976 AND 1980

		^{144}Ce + ^{144}Pr	^{125}Sb	^{106}Ru + ^{106}Rh	^{137}Cs
East Saint-Vaast	8.3.77	1.64	0.23	3.0	1.12
	16.10.78	0.81	0.21	2.4	0.60
	10.5.79	0.95	0.29	8.3	0.83
	10.1.80	0.56	0.23	5.2	0.67
East Ouistreham	17.2.77	1.16	0.16	4.0	1.5
	20.7.79	1.03	0.22	10.3	0.92
	8.1.80	0.7	0.30	12.7	0.93
West Granville	8.7.76	9.7	0.26	10.9	0.69
	18.2.77	8.2	0.24	12.4	0.77
	24.3.78	4.7	0.24	9.6	0.62
	21.6.78	4.4	0.23	8.8	0.65
	26.3.79	2.1		11.7	
	10.9.79	1.5		7.9	
	16.11.79	1.1	0.18	7.2	0.46
	4.3.80	0.8		7.2	0.37

In such cases, soluble ^{106}Ru in the suspended solids or sediments will be retained by adsorption of insoluble compounds at a rate dependent on the residence time and equilibrium constants.

2.4. Study of ^{106}Ru/^{144}Ce ratios

Tables I to III and Fig. 2 give only mean values observed at several reference sites (10–20 samples per site). A logarithmic scale graph with ^{106}Ru and ^{144}Ce as variables was chosen to take into account all other coastal points and all samples taken.

The distributions observed in February 1977 in the Normandy-Brittany Gulf (Granville, Le Vivier, mouth of the Fremur and Arguenon Rivers, Bay of Frênaye, mouth of the Trieux River), the Bay of the Seine (Port-en-Bessin, Courseulles,

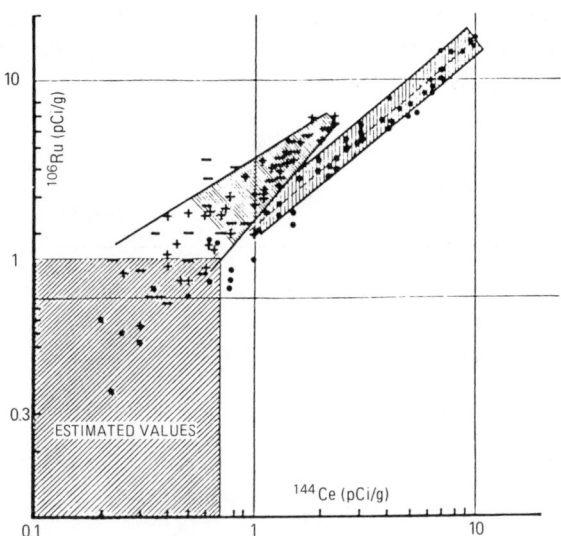

FIG.3. Relation between ^{106}Ru and ^{144}Ce in western (●) and eastern [(+) Seine Bay, (−) North Sea] sediments in February 1977.
▨ *estimated values.*

Orne Canal, Orne River, Seine Estuary, mouths of the Dives and Touques Rivers) show that the points representative of sediments collected to the east and west of the La Hague plant discharge structure (Fig. 3) are grouped in two separate clusters, as follows:

(a) all samples collected in the Bay of Mont-Saint-Michel, as well as those taken further to the west between La Rance and the Trieux River (150 km coastal area) distribute around a line with a slope of one characterized by a constant ^{106}Ru/^{144}Ce ratio of 1.5, with a correlation factor greater than 98%;

(b) distribution to the east is less uniform.

The gap between the two clusters reflects the reconcentration to the west of the two radionuclides considered, with a predominance of ^{144}Ce in relation to ^{106}Ru. The linear relationships in the Normandy-Brittany Gulf are verified at surface and subsurface levels between 1976 and 1978 (see Table IV). This condition is not changed by the previously mentioned fine silt deposits at Cancale and Granville in March 1978, which caused an increase in activity levels, because the ^{106}Ru/^{144}Ce ratios are identical.

The cause of this phenomenon, observed with northern winds, was identified as a shifting of sediments from the subtidal zone in the bay. The increase in activity was due to silt particle return to suspended form and

TABLE IV.   RATIOS OF ^{106}Ru/^{144}Ce IN VERTICAL SEDIMENTARY PROFILES
FROM GRANVILLE, LE VIVIER, CANCALE (MONT-SAINT-MICHEL BAY)
AND FERMANVILLE (NW COTENTIN)

	Granville 21/2/77	Le Vivier 24/3/78	Cancale 3/7/78	Fermanville 8/3/77
0 − 2.5	1.39	2.34	1.69	1.04
2.5 − 5	1.52	2.00	1.80	1.08
5 − 7.5	1.13	2.11	1.79	0.91
7.5 − 10	1.30	1.72	−	0.59
10 − 12.5	1.40	1.56	1.19	0.51
12.5 − 15	1.23	1.40		0.51
15 − 17.5	1.16	2.62		0.49
17.5 − 20.0	1.29	2.14		1
20 − 25	1.58	2.15		
25 − 30	0.90	2.04		
30 − 35	2.51	1.69		
35 − 40	2.35			
40 − 45	2.35			
45 − 50	1.15			
etc.				

temporary deposit in the Bay of Cancale. Significantly, this silt deposited at
surface level during May and July exhibits an extractable organic carbon content
(addition of sodium pyrophosphate) of 70 to 100%, which is quite different
from surface level observations (Fig. 4) in 1977. The existence of high extractable
carbon contents moreover indicates the presence of loose silt deposits.

Changes in the ^{106}Ru/^{144}Ce ratios recorded at La Hague, Fermanville,
Granville and Cancale (Fig. 5) show that a non versatile modification in the
composition of the radioactive wastes at the time of release was entirely observed
at least one year later in the Bay of Mont-Saint-Michel.

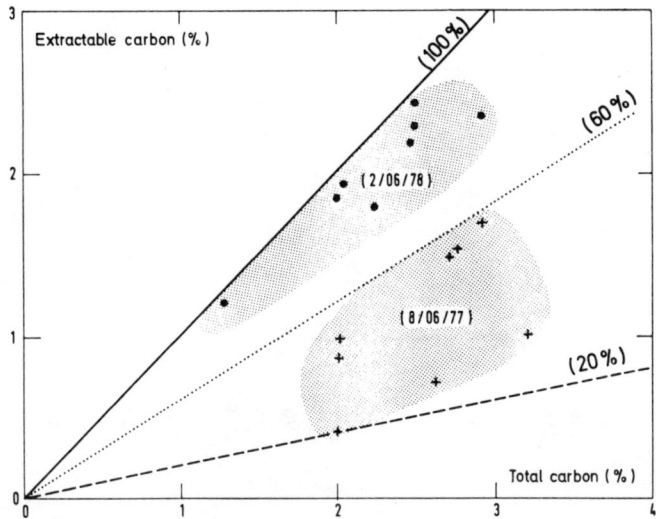

FIG.4. Comparison between total and extractable carbon from two sampling periods in Mont-Saint-Michel Bay (after A. SAAS, P. GUEGUENIAT, to be published).

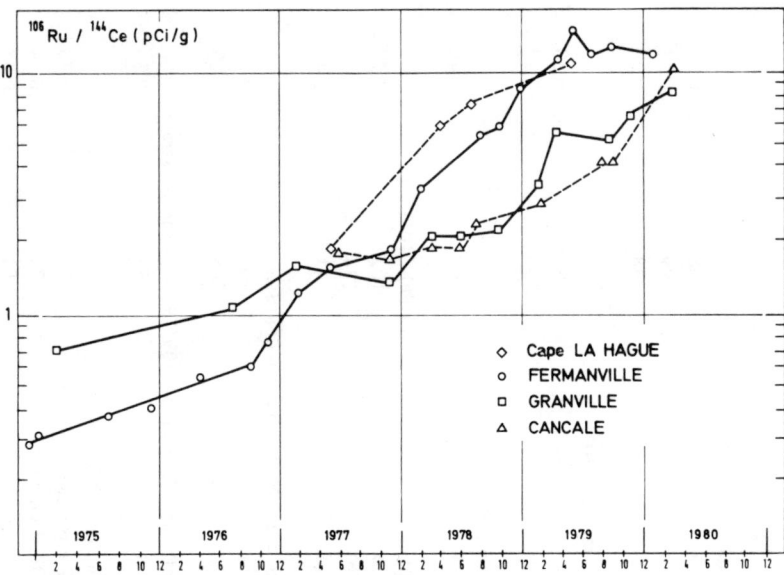

FIG.5. Ratios of $^{106}Ru/^{144}Ce$ in sediments from La Hague, Fermanville, Granville, Cancale.

TABLE V. LEVELS (pCi/kg DRY) OF ^{239}Pu, ^{106}Ru ($+$ ^{106}Rh) IN MUDDY COASTAL SEDIMENTS FROM THE CHANNEL. EVOLUTION OF THE RATIOS ^{238}Pu/^{239,240}Pu; ^{106}Ru/^{144}Ce

	^{239}Pu	^{238}Pu/^{239,240}Pu	^{106}Ru + ^{106}Rh	^{106}Ru + ^{106}Rh/^{239}Pu	^{106}Ru/^{144}Ce
Fermanville 27/04/78					
(a)	68	0.50	6 600	97	5.5
(b)	99	0.31	12 500	126	3.4
(c)	98	0.23	21 400	218	4.3
Fermanville 12/04/79					
(a)	97	0.30	28 000	287	10.7
(b)	350	0.37	160 000	457	10.9
Fermanville 27/07/79					
(a)	242	0.40	65 000	268	11.8
Mont-Saint-Michel Bay (Granville) 24/03/78					
(a)	102	0.22	9 800	96	2.03
(b)	134	0.25	13 100	98	2.09
(c)	41	0.49	16 100	393	2.07
Mont-Saint-Michel Bay 21/08/79					
(a) Cancale	103	0.22	6 000	58	4.1
(b) Le Vivier	288	0.09	7 800	27	4.9
Mont-Saint-Michel Bay 7/09/79 (Vivier lower strand)					
(a)	53	0.35	2 000	40	
Seine Bay 20/07/79					
Courseulles	78	0.33	11 000	141	10.1
Quistreham	77	0.34	10 650	138	9.4
Honfleur	71	0.38	6 140	86	8.6
Le Havre	74	0.24	9 000	122	6.3
Seine	68	0.41	7 800	115	5.8

3.   PLUTONIUM DISTRIBUTION IN ENGLISH CHANNEL SEDIMENTS

As for cerium, maximum plutonium releases from the La Hague plant occurred during the winter of 1974—1975 (15 Ci) and then diminished (about 6 Ci per year). Between 1974 and 1979 the $^{238}Pu/^{239}Pu$ ratios increased significantly from 0.14 to 0.50. In the La Hague coastal waters, maximum ^{239}Pu contents were observed in December 1974 at concentrations of 50 and 100 fCi/ltr with 35 to 65% in particulate form and subsequently below 10 fCi/ltr.

Plutonium has a long half-life and exhibits properties comparable to ^{144}Ce regarding its particulate form in sea water and distribution in wastes. It is thus possible to validate the hypotheses proposed concerning transport of ^{144}Ce to the west, especially since samples can be monitored by observing changes in the $^{238}Pu/^{239}Pu$ ratio.

A limited number of samples selected on the basis of ^{106}Ru and ^{144}Ce data were initially analysed as follows:
(i)   3 sediment samples from Fermanville on April 27, 1978, determining radio-nuclide activity observed (^{106}Ru and ^{144}Ce mean and extreme values),
(ii)   2 sediment samples from Fermanville on April 12, 1979, marking the highest ^{106}Ru activity periods; selection of samples representing extreme activity levels,
(iii) the highest activity sample at Fermanville on July 24, 1979,
(iv)   3 samples (mean and extreme   activity levels) from the Bay of Mont-Saint-Michel at Granville on March 24, 1978, corresponding to deposit of loose silt with traces of ^{144}Ce released in 1974—1975,
(v)   one sample on August 21, 1979 from the port of Le Vivier entrance channel and one sample from the port of Cancale (Bay of Mont-Saint-Michel),
(vi)   one foreshore sample at Le Vivier on September 7, 1979,
(vii) the most active samples on July 20, 1979 (one per site), collected in the Bay of the Seine at Ouistreham, Courseulles, Honfleur, Le Havre and the Seine Channel.

Observations pertaining to plutonium distribution and the radionuclides contained in the above samples are presented in Table V. The samples from Fermanville contain from 100 to 300 pCi/kg in 1978 and 1979. Moreover, for ^{106}Ru and ^{106}Rh, the highest activity levels are also present in the same samples with $^{238}Pu/^{239, 240}Pu$ ratios near 0.40, which correspond to isotopic ratios of the present wastes.

Data obtained in 1979 for the Bay of the Seine are highly uniform for ^{239}Pu (68—77 pCi/g) with $^{238}Pu/^{239, 240}Pu$ ratios of 0.33 to 0.34 on the coast (two samples) and of 0.24 to 0.41 for the estuarine portion of the Seine (three samples). The Cotentin/Bay of the Seine activity ratios are only three for plutonium versus nine for ^{106}Ru in the same samples.

TABLE VI. DESORPTION BY HNO$_3$ OF PLUTONIUM RETAINED BY SEDIMENTS CONTAMINATED EITHER IN SITU (FROM FERMANVILLE — [239]Pu) OR EXPERIMENTALLY (SEDIMENTS WITH OR WITHOUT HYDROCARBONS). RESULTS ARE EXPRESSED AS PERCENTAGES

	pH2	pH1	0.5N	1N	2N	4N	5N	6N	8N
Fermanville sediment (contaminated in situ)									
	10	42	53	71	51	57			64
Sample without hydrocarbon (contaminated experimentally)									
(1)	12	27	25	26	34		56	72	87
(2)	13	32	61	54	65		81	82	84
Sample with hydrocarbon (40 000 ppm)									
	0	11	23	22	19		20		88

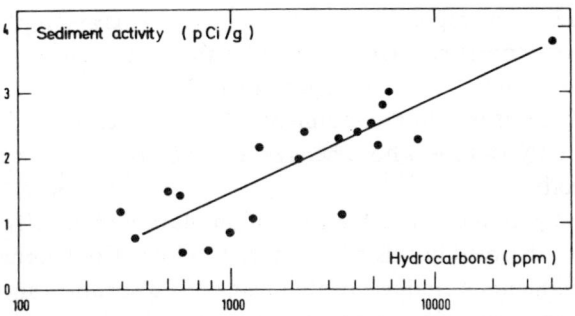

FIG.6. Activity of ^{144}Ce in sediments versus hydrocarbon concentration pollution in the western Finistère from July 1978 to February 1979.

In the Bay of Mont-Saint-Michel samples analysed, the reconcentration observed for ^{144}Ce also seems to occur, to a lesser extent, for plutonium. This can essentially be attributed to the releases during the winter of 1974—1975 because the most active sample (288 pCi/kg) presents a ^{238}Pu/$^{239, 240}$Pu ratio of 0.09, which is characteristic of the former releases. Similarly, for three other Bay of Mont-Saint-Michel sediment samples (out of six investigated) having activities ranging from 103 to 134 pCi/kg, these same ratios (0.2—0.25) show the distinct influence of the 1974—1975 releases. In contrast, some samples from this same region reflect the present composition of La Hague releases with low ^{239}Pu levels (41—53 pCi/kg), not as high as in the Bay of the Seine. Moreover, for the Bay of Mont-Saint-Michel sediments, there is no relationship between the ^{239}Pu and ^{106}Ru or ^{144}Ce activity levels.

4.   ARTIFICIAL RADIONUCLIDE TRANSPORT IN OIL-CONTAMINATED SEDIMENTS

In March 1978 the coastal area investigated between Lannion and Brest since 1976 (to determine effects of the 18th-22nd Chinese nuclear weapons tests) was seriously contaminated by the Amoco Cadiz oil spill. Radioanalysis of contaminated samples showed a sharp ^{144}Ce increase in sediments when the hydrocarbon content exceeded 1000 ppm (Fig. 6). Only one sediment sample had a hydrocarbon greater than 10 000 ppm. The hydrocarbon content of this sample was 40 000 ppm and its ^{144}Ce content was 3.8 pCi/g, which places it outside the activity levels observed in the regions subject only to nuclear fallout [6]. Taking into account the particulate nature of ^{144}Ce in sea water, it was assumed that the hydrocarbons absorbed particles suspended in the sea water at surface and subsurface levels, thus resulting in the displacement of a large water mass followed by more or less rapid deposit on the coast. This assumption regarding subsurface action is consistent with the rapid water column diffusion of hydrocarbons observed by Marchand and Caprais in 1978. Of the three artificial radionuclides regularly measured in the sediments (^{144}Ce, ^{106}Ru and ^{137}Cs), it appears that ^{144}Ce is the only isotope whose behaviour depends on the degree of hydrocarbon contamination.

In situ investigation of the effects of oil on plutonium is difficult because the activity levels observed in this region were too low. Contaminated and non-contaminated sediments were investigated using ^{237}Pu sorption and desorption techniques. The plutonium (valence + 6) adsorption rates were of no particular interest, since after 24 hours the distribution coefficients were $0.6-4 \times 10^5$ and the lowest values were observed for contaminated sediments.

The HNO$_3$ solution acidity desorption studies were, however, of greater interest. Table VI shows the data for non-oil-contaminated sediments with desorptions of 30 to 40% at pH 1 and 56 to 80% in 5N HNO$_3$ environment; with the oil-contaminated sediments the desorption is more difficult: only 10% at pH 1, 20% in 5N HNO$_3$, possibly because of plutonium-hydrocarbon binding. The existence of plutonium observed experimentally to be easily desorbable from sediments in natural conditions was confirmed by the field investigations at La Hague, with sediment samples from Fermanville and at Windscale [9].

5.   CONCLUSION

The investigations of gamma-emitting radionuclide distribution in coastal surface sediments in the English Channel between 1976 and the beginning of 1980 demonstrate the extent and complexity of transport to the west of La Hague. The initial data for similar studies in 1978–1979 pertaining to plutonium show activities of less than 400 pCi/kg, which is very low compared with the Irish Sea with activities of 300 000 pCi/kg in the immediate vicinity of the Windscale plant and 15 000 to 20 000 pCi/kg at a distance of 15 km. They also confirm the significance of transport to the west.

Taking into account the larger releases from the La Hague plant during the winter of 1974–1975, compared with preceding and subsequent periods, the high proportion of particulates in 1974–1975 and the ^{238}Pu/$^{239,\,240}$Pu ratio increase (0.13 to 0.50) between 1974 and 1980, radionuclide measurements confirmed most of the assumptions made concerning gamma radionuclides, namely:

(a) preferential reconcentration of particulates in the Bay of Mont-Saint-Michel and, more generally, in all of the Normandy-Brittany Gulf,

(b) retention of activity discharged in La Hague during the winter of 1974–1975 in the Bay of Mont-Saint-Michel surface sediments collected in 1978–1979, despite 1 500 000 m^3 of silt deposited every year in this region.

It has not yet been possible to conduct an investigation similar to the [144]Ce field study for the Amoco Cadiz-contaminated sediments in a zone of the Normandy-Brittany Gulf unaffected by the releases. A significant increase in coastal sediment activity of [144]Ce was observed in this region when contamination exceeded 1000 ppm. This was attributed to displacement of the suspended particulates and cerium associated with the oil spill, followed by deposit on the coast. Experimental study of plutonium desorption using HNO_3 solutions of increasing acidity revealed another aspect of the effects of hydrocarbons on radionuclide transport. In addition to the mechanical action observed for [144]Ce, there is also a chemical effect in that, when hydrocarbons are present, plutonium becomes much more difficult to desorb than in natural conditions due apparently to direct bonding with hydrocarbons.

REFERENCES

[1]   JEFFERIES, D., PRESTON, A., STEELE, A., Distribution of caesium-137 in British coastal waters, Mar. Pollut. Bull. 4 8 (1973) 118.

[2]   LIVINGSTON, H., BOWEN, V., "Americium in the marine environment-relationships to plutonium", in Environmental Toxicity of Aquatic Radionuclides-Models and Mechanisms (MILLER, M.N.N., STANNARD, J.N., Eds), Ann Arbor Science Publishers, Ann Arbor, Michigan (1976) 107.

[3]   MURRAY, C.N., KAUTSKY, H., Plutonium and americium activities in the North Sea and German coastal regions, Estuar. Coast. Mar. Sci. 5 3 (1977) 319.

[4]   KAUTSKY, H., "The Norht Sea region taken as an example for the behaviour of artificial radioisotopes in nearshore areas", in Proc. of the Third NEA Seminar, Tokyo, 1–5 October 1979.

[5]   GUEGUENIAT, P., AUFFRET, J.P., BARON, Y., Evolution de la radioactivité artificielle gamma dans des sédiments littoraux de la Manche pendant les années 1976–1977–1978, Oceanol. Acta 2 2 (1979).

[6]   GUEGUENIAT, P., AUFFRET, J.P., KHALANSKI, M., DUPONT, J.P., "Evolution dans les sédiments marins de radionucléides d'origine industrielle en présence d'hydrocarbures", 2nd Int. Symp. on Radioecology, Cadarache 19–22 June 1979.

[7]   GUEGUENIAT, P., GANDON, R., LUCAS, Y., "Determination of radionuclides of Ce, Co, Fe, Zn and Zr in sea water by preconcentration of colloidal manganese dioxide", Reference Methods for Marine Radioactivity Studies II, Technical Reports Series No. 169, IAEA, Vienna (1975) 137.

[8]   GUEGUENIAT, P., Comportement physico-chimique du ruthénium de fission dans le milieu marin, Rep. CEA-R-4644 (1975).

[9]   HARVEY, B.R., "Interstitial water studies on Irish Sea sediments and their relevance to the fate of transuranic nuclides in the marine environment", these Proceedings.

INTERSTITIAL WATER STUDIES ON IRISH SEA SEDIMENTS AND THEIR RELEVANCE TO THE FATE OF TRANSURANIC NUCLIDES IN THE MARINE ENVIRONMENT

B.R. HARVEY
Ministry of Agriculture, Fisheries and Food,
Directorate of Fisheries Research,
Fisheries Radiobiological Laboratory,
Lowestoft, Suffolk,
United Kingdom

Abstract

INTERSTITIAL WATER STUDIES ON IRISH SEA SEDIMENTS AND THEIR RELEVANCE TO THE FATE OF TRANSURANIC NUCLIDES IN THE MARINE ENVIRONMENT.
    This paper describes the physico-chemical conditions existing in the interstitial waters of sediments in contaminated areas of the Irish Sea, which provide valuable information on the sedimentary environment into which radioactive waste products become incorporated. It is recommended that these measurements be made in areas where transuranic behaviour can be determined, which then would allow useful predictions to be made concerning the possible behaviour of transuranics in other, uncontaminated, environments, if these environments can be physico-chemically correlated in the same way.

1.   INTRODUCTION

One of the important considerations in determining the long-term effect of artifically produced radionuclides in the marine environment is the possible return of such nuclides to the water column after burial in bottom sediments.

Work on the environmental behaviour of transuranic elements has been somewhat difficult because of the extremely small quantities of these elements present, most of which come from atmospheric fallout. The Irish Sea, however, provides an almost unique opportunity to study these elements in at least one type of marine environment, because it receives low-level aqueous radioactive wastes discharged from the British Nuclear Fuels reprocessing plant at Windscale in Cumbria.

Sediments form a natural sink for many polluting substances including the transuranic elements, but a study of the extent to which these and other elements partition between the water and sediment phases in the sea (Table I) shows that there are marked differences between elements. Furthermore, it appears that a relationship exists between the oxidation state of the element present and the

TABLE I. DISTRIBUTION COEFFICIENTS FOR TRANSURANIUM AND
OTHER ELEMENTS BETWEEN SOLID AND LIQUID PHASES IN A MARINE
ENVIRONMENT

Element	$K_d = \dfrac{\text{concentration in sediment}}{\text{concentration in water}}$	Valency state	References
^{232}Th	$> 10^7$	IV	From Burton [12]
U	$\approx 10^3$	VI	From Burton [12]
^{237}Np	$\approx 5 \times 10^3$	V	Present work
$^{239/240}$Pu	3.5×10^5 2.5×10^6 5×10^3	Total IV V or VI	Nelson and Lovett [3]
^{241}Am	2.3×10^6	III	Pentreath et al. [11]
Nd natural	2×10^6 to 1×10^7	III	From Mason [13]

distribution coefficient of the element with respect to the aqueous and solid phases
of the system.

The distribution coefficient in Table I is defined as:

$$K_d = \frac{\text{concentration in the sediment}}{\text{concentration in the water}}$$

The naturally occurring radioelements thorium and uranium provide an
illustration. Thorium is invariably present in the tetravalent state (Th (IV)) and is
extremely insoluble in sea water; it shows an overwhelming affinity for sediments.
On the other hand, uranium is generally considered to be present in sea water as the
highly soluble UO_2^{2+} carbonate complex and shows a very low K_d. (This is reflected

in the fact that sea water usually contains in excess of $3\mu g \cdot l^{-1}$ of uranium.) In strongly reducing environments, however, uranium is reduced to the tetravalent state and it is generally supposed that uranium ores were laid down under strongly reducing environmental conditions [1].

With the transuranium elements it can be seen from Table I that the elements americium and curium, which normally exhibit only tri-valency, are similar to the lanthanide elements. Indeed, neodymium has been used by various workers as a useful analogue for americium in experimental work [2]. The somewhat lower overall K_d given for plutonium has been shown [3], for the Irish Sea at least, to be a composite of the distribution coefficients for two oxidation states, Pu(IV) and Pu(V) or (VI). The even lower K_d shown for neptunium is in keeping with the generally accepted view that this element exists in an oxidized sea-water environment as the highly soluble NpO_2^+ ion (pentavalent). It seems reasonable to suppose therefore that environmental changes which bring about changes in the oxidation/reduction potential of the system will also markedly alter the affinity shown by some transuranics for the solid or liquid phase.

Sediments can normally be expected to display more reducing conditions than those found in the overlying sea water, but it is the extent to which these reducing conditions develop that are important in a consideration of post-depositional mobility of elements within the bed material. This phenomenon of migration within the sedimentary system has an important bearing on the ultimate fate of elements in the environment.

2.   EXPERIMENTAL

2.1. Leaching of Irish Sea sediments

Experiments were carried out to investigate the extent to which transuranic elements associated with the surface sedimentary material might be in some sort of equilibrium condition with the overlying water in the Irish Sea. Despite the high affinity of many of them for the particulate phase in the Irish Sea system it was thought unlikely, bearing in mind the nature of the discharge source, that they would display such refractory characteristics as has been reported to exist at some sites of discharge [4].

Various amounts of sediment from the eastern Irish Sea area, of known plutonium and americium content, were placed in 1200 ml polythene bottles. To these were added 1-litre samples of filtered sea water (0.22 μm Millipore) from the North Sea having a very low plutonium and americium content. The bottles were rolled for various time periods to mix the contents, after which the contents were carefully filtered again, making sure that no sediment particles were allowed to remain with the filtered water. After spiking with suitable plutonium and

TABLE II(a). LEACHING OF PLUTONIUM AND AMERICIUM FROM IRISH SEA SEDIMENTS WITH 'CLEAN' NORTH SEA WATER (1 LITRE)

Leaching time (h)	Amount of activity leached (mBq)							
	0.05 g Sediment		0.5 g Sediment		5 g Sediment		20 g Sediment	
	Pu	Am	Pu	Am	Pu	Am	Pu	Am
6	3.3	3.5	7.7	4.1	–	–	–	–
16	3.2	3.8	7.5	4.3	–	–	–	–
24	3.0	3.6	10.2	4.2	15.2	3.3	18.5	4.4
48	–	–	–	–	21.8	3.7	25.5	4.8
96	–	–	–	–	24.8	3.3	20.0	4.4

Initial activity on sediment Pu   2.5 Bq·g^{-1} Am   1.9 Bq·g^{-1}
in water   0.3 mBq·l^{-1} 0.4 mBq·l^{-1}

TABLE II(b). REPEATED LEACHING OF PLUTONIUM AND AMERICIUM FROM THE SAME SAMPLE OF SEDIMENT WITH 'CLEAN' SEA WATER
(0.05 g sediment with 1-litre samples of water)

Leaching time (h)	Amount of activity leached (mBq)			
	1st Leach		2nd Leach	
	Pu	Am	Pu	Am
6	3.0	3.5	2.1	4.0
16	3.3	3.8	2.1	4.1
24	3.0	3.6	1.6	2.5

Note: Standard deviation on the measurements approximately ± 10%.

americium recovery tracers the water samples were then analysed for $^{239/240}$Pu and ^{241}Am. The results of these experiments are shown in Tables II(a) and II(b) and it is clear that equilibrium is approached in a comparatively short period of time. Table II(a) shows that both plutonium and americium are leached from the sediment. For americium the amount partitioning into the water phase appears to be independent of the amount of sediment present, at least with the range of sediment weights taken. This is in keeping with the expected distribution coefficient for americium. Presumably, if a sufficiently small amount of sediment were to be used, a limiting point would be reached. For plutonium, on the other hand, some increased leaching occurs as the weight of sediment present is increased. This almost certainly reflects the presence of a small amount (say 1 or 2%) of a higher oxidation state which has a much lower K_d value than the bulk of the plutonium (see Table I).

Table II(b) shows the effect of repeating the leaching on the same portion of sediment with a fresh 1-litre sample of clean water. It illustrates that the activity being removed cannot be accounted for (as might be inferred from a single leaching experiment) by assuming that plutonium and americium contained in the interstitial water of the sediment was the source of the plutonium and americium found to be extracted. For americium the same concentration occurs in both the first and the second leaches. However, it can be seen that somewhat less plutonium is present in the second leach, suggesting again that a small portion of a more soluble plutonium species was present in the original sample.

Table III is derived from the same experimental data as given in Tables II(a) and II(b) but shows the concentration in the aqueous extract versus the total

TABLE III. LEACHING OF PLUTONIUM AND AMERICIUM FROM IRISH SEA SEDIMENTS WITH 'CLEAN' NORTH SEA WATER (1 LITRE)

	0.05 g Sediment		0.5 g Sediment		5 g Sediment		20 g Sediment	
	Pu	Am	Pu	Am	Pu	Am	Pu	Am
Total activity on sediment (mBq)	125	95	1250	950	12500	9500	50000	19000
Concentration in water after 24 h (mBq · l^{-1})	3.0	3.5	10.2	4.2	15.2	3.7	18.5	4.4
Apparent K_d	8.3×10^5	5.4×10^5	2.5×10^5	4.5×10^5	1.6×10^5	5.1×10^5	1.4×10^5	4.3×10^5

$$K_d = \frac{\text{concentration on sediment}}{\text{concentration in water}}$$

amount of plutonium and americium present in the sample of sediment used. The apparent distribution coefficients for the two elements shown at the bottom of the table again suggest the presence of two components for plutonium with differing distribution coefficients whilst americium shows no such variability.

These simple leaching experiments are shown here primarily to demonstrate that in the Irish Sea we are dealing with a dynamic system where changes in one compartment of the system bring about adjustments in others. We may thus conclude that exchange will occur between the upper layers of the bed and the overlying water and, furthermore, that these exchanges will be comparatively rapid processes and that they will reflect the distribution coefficients for the respective elements. However, if conditions alter so that the oxidation state of the element changes, then it is obvious that the distribution between the aqueous and solid phases will change. This type of change can occur within the sediment bed.

2.2. Interstitial water studies

It is clear that the pore fluids in the interstices of bottom sediments hold the key to post-depositional mobility of substances which become buried within sediments. The upward migration of manganese [5] has probably been documented better than that of any other element but many have been investigated [1,5]. Ideally, of course, it would be desirable to make direct measurements in order to study the mobility of transuranics, but even in the Irish Sea the concentration of these elements in the pore fluids makes this difficult and very time consuming. To study the oxidation state of the elements is even more difficult due to the much larger samples of pore fluids required for such separate determinations.

At present, therefore, useful predictions have to be made about the post-depositional behaviour of the transuranics by measuring other parameters in the pore fluids from sediments in the appropriate area, and then combining these data with the known characteristics of the transuranium elements to assess their potential for migration. This approach can never be a substitute for direct measurements, but it does provide a useful framework within which to predict the behaviour of the various elements of interest.

Experimental work was carried out during research cruises on the MAFF Fisheries Research Vessel CIROLANA. Cores up to 1 metre in length were obtained using a 4-inch gravity corer. These were extruded and cut immediately; the outside portion of each section was discarded to guard against contamination with surface water and precautions were taken to exclude air as far as possible. Sections of the core were transferred immediately to a nitrogen-filled glove box in which they were squeezed by hydraulic pressure to expel the pore-fluid. The squeezing rig was fitted with a 0.45 μm Millipore filter on the outlet to ensure the removal of all sediment particles. An in-line platinum redox probe provided a continuous record of the E_h value of the expelled fluid. The E_h value dropped rapidly during the

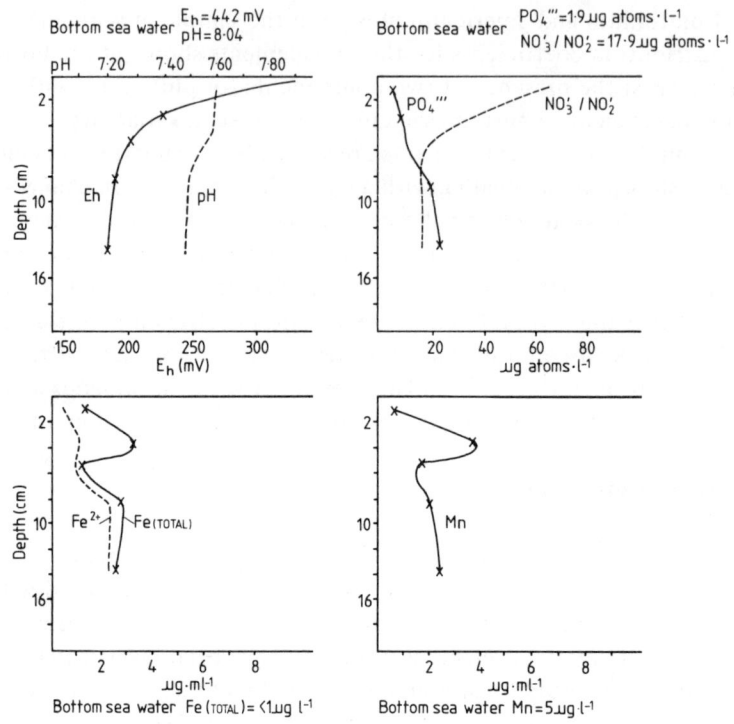

FIG.1. Interstitial waters from an Irish Sea sediment (March 1979).

early period of squeezing as the oxygen trapped in the filter paper and other parts of the apparatus was expelled [6]. The sample of interstitial water on which further analyses were carried out was collected when the E_h value had dropped to a constant value as shown.

Figure 1 summarizes the measurements made on three cores in the eastern Irish Sea area. It can be seen that the E_h and pH values both drop rapidly away from the sediment surface. The E_h value in these sediments tends to stabilize at between 100 and 200 mV (on the hydrogen scale). Note that the dissolved iron builds up rapidly and this is largely in the ferrous state, though organically complexed ferric iron is also present. The presence of nitrate/nitrite, which is almost certainly produced by bacterial action within the bed, tends to hold the redox slightly higher in the upper parts of the bed, but eventually this gives way to the ferric/ferrous system which is then the dominant influence in the redox potential for some considerable depth. Virtually no sulphide can be detected in these sediments (though there are isolated areas in the Irish Sea where there are strongly reducing conditions). In general, therefore, we are dealing with a sediment-water interface which is well oxygenated, and below this a large area, in terms of depth, where free oxygen is limited but where alternative electron acceptors are being utilized to effect the oxidation of organic matter. Conditions in these sediments can be classed as mildly reducing with an E_h value which ranges from 100 to 200 mV. These are the conditions in which we must consider the behaviour of transuranic elements in the Irish Sea sedimentary environment.

3. DISCUSSION

In considering the post-depositional fate of transuranic nuclides in sediments it is worth looking to see what can be learnt from the naturally occurring α-emitting radionuclides. Bonatti et al. [1] concluded that there is no evidence for the post-depositional migration of thorium in marine sediments, at least during early diagenesis. Uranium, on the other hand, is known to accumulate where strong reducing conditions exist because it is reduced to the tetravalent state. This is not to say, however, that an ascending migration of uranium cannot take place in the less reducing parts of a sediment [7,8] and this is, of course, known to occur. One further point is worth noting about naturally occurring elements. It has been shown that lanthanide elements accumulate in the less reducing zones of sediments because they associate with phosphate. Phosphates are released from strongly reducing sedimentary environments due to the conversion of iron and calcium phosphates to FeS and $CaCO_3$ (chiefly by bacterial action) and, thus, phosphates concentrate in the mildly reducing zones along with such cations as the lanthanides and manganese.

Turning now to the transuranic elements, it can reasonably be assumed that both americium and curium will remain firmly fixed as trivalent forms in sediments hosted with phosphates and the lanthanides. There would seem therefore to be a negligible tendency for these elements to solubilize from mildly reducing sediments. From a comparison of the redox and other measurements made on Irish Sea sediments with published stability relationships for dissolved plutonium species in water [9], it can be calculated that higher oxidation states of plutonium (i.e. (V) and (VI)), which have been shown to exist in oxidized sea water, will be reduced below an E_h value of, say, about 250 mV. If these assumptions are correct, then the post-depositional mobility of plutonium in sediments, as a result of redox changes at least, will be severely restricted. There remains, however, the possibility that organic complexing agents may in some situations increase the solubility of the lower valency states of plutonium.

Neptunium is generally considered to exist in oxidized sea water as the highly soluble pentavalent neptunyl ion and, as has been shown earlier, it displays a low affinity for sedimentary material in a manner similar to the uranyl ion. However, the redox potential for the reduction of Np(V) to Np(IV) at the pH value existing in

Irish Sea interstitial waters is probably between 100 to 150 mV. In these mildly reducing sediments, therefore, neptunium may be slow to reduce. Ground water studies [10] showed that, whilst various basalts and granites were effective in reducing neptunium to the tetravalent state, some unweathered shales were not. Work is at present in progress to investigate the reduction of neptunium in Irish Sea sediments, because of the importance of ^{237}Np as a component of radioactive wastes and because of its generation within sedimentary systems by the decay of its parent ^{241}Am and grandparent ^{241}Pu.

REFERENCES

[1]   BONATTI, E. et al., Geochim. Cosmochim. Acta **35** (1971) 189.
[2]   WEIMER, W. et al., Proc. 3rd Int. Conf. Nuclear Methods in Environmental and Energy Research, CONF-77 1072, Columbia (1977) 472; Am. Nucl. Soc. (1979).
[3]   NELSON, D., LOVETT, M.B., Nature **276** (1978) 599.
[4]   PILLAI, K., MATTHEW, E., GANGULY, A., Studies on Plutonium in the Marine Environment, Bhabha Atomic Research Centre, Report BARC/1-462 (1977) 141.
[5]   DUCHART, P., CALVERT, S., PRICE, N., Limnol. Oceanogr. **18** (1973) 605.
[6]   TROUP, B., BRICKER, O., BRAY, J., Nature **249** (1974) 237.
[7]   KOLODNY, Y., KAPLAN, I., Geochim. Cosmochim. Acta **34** (1970) 3.
[8]   KU, T., J. Geophys. Res. **70** (1965) 3457.
[9]   POLZER, W., in Safety in Plutonium Handling Facilities (Proc. Symp. Rocky Flats, 1971), USAEC CONF-710401 (1971) 411.
[10] BONDIETTI, E., FRANCIS, C., Science, N.Y. **203** (1979) 1337.
[11] PENTREATH, R.J. et al., Proc. 3rd NEA Seminar on Marine Radioecology (Tokyo, Oct. 1979) OECD, Paris (1980) 291.
[12] BURTON, J.D., in Chemical Oceanography (RILEY, J.P., SKIRROW, G., Eds) **3**, 2nd Edn., Academic Press, London and N.Y. (1975) 91.
[13] MASON, B., Principles of Geochemistry, John Wiley and Sons, N.Y. and London (1966) 181 and 196.

A PROPOSED APPROACH TO THE STUDY
OF RADIONUCLIDE GEOCHEMISTRY
IN INTERTIDAL SEDIMENTS*

A.B. MacKENZIE
Scottish Universities Research
    and Reactor Centre,
East Kilbride, Glasgow,
United Kingdom

Abstract

A PROPOSED APPROACH TO THE STUDY OF RADIONUCLIDE GEOCHEMISTRY IN
INTERTIDAL SEDIMENTS.
    Information on the physical and chemical partitioning of radionuclides within marine
sediments is important in attempting to formulate a description of their marine geochemistry.
This is particularly applicable in the case of elements such as plutonium which can exist in a
number of oxidation states with different solubility characteristics. In coastal and estuarine
locations intertidal sediments can represent an important geochemical environment and these
deposits often exhibit inhomogeneous physical and chemical characteristics resulting in major
variations in radionuclide partitioning within the sediment column. A study of natural and
man-made radionuclides in intertidal sediments from south-west Scotland has recently been
started at SURRC and, while only bulk sediment samples are being analysed at present, initial
results indicate that a more detailed study will be required to rationalize the observed radionuclide distribution patterns. It is therefore proposed to extend the present study to a more
broadly based geochemical investigation involving radiochemical analysis of selected physical
and chemical fractions of the sediment in conjunction with instrumental analytical techniques.

The Clyde Sea Area, located in south-west Scotland, provides an important
amenity for nearby industrial areas, being used for navigation, fishing, industrial
purposes, recreation and waste disposal activities [1]. The area receives significant
inputs of both radioactive and stable pollutants and, in such an intensively utilized
locality, it is important to determine the geochemical behaviour of these pollutants.
Concentrations of man-made radionuclides in sea water and marine sediments in
this area are almost entirely dominated by nuclides discharged in liquid effluent
from the BNFL reprocessing plant at Windscale, north-west England. Major
sources of stable pollutants are in the industrial areas of the Clyde Estuary and
northern sections of the Firth of Clyde, with smaller contributions from atmospheric
input and mixing with previously contaminated Irish Sea water. A considerable

---

* This work was partly financially supported by the International Atomic Energy Agency
under Research Contract No. 2409/RB.

257

TABLE I.   RADIONUCLIDES BEING ANALYSED IN
INTERTIDAL SEDIMENTS

Nuclide	Occurrence	Analytical method
^{40}K	Natural	Gamma spectroscopy
^{106}Ru	Man-made	Gamma spectroscopy
^{134}Cs	Man-made	Gamma spectroscopy
^{137}Cs	Man-made	Gamma spectroscopy
^{144}Ce	Man-made	Gamma spectroscopy
^{210}Bi	Natural	Radiochemical separation and beta counting
^{210}Po	Natural	Radiochemical separation and alpha spectroscopy
^{226}Ra	Natural	Separation and alpha counting of ^{222}Rn daughter
^{238}Pu	Man-made	Radiochemical separation and alpha spectroscopy
^{239,240}Pu	Man-made	Radiochemical separation and alpha spectroscopy

research effort has been devoted to the study of both radioactive and non-radioactive marine pollution in this area in recent years [2—7]. Some of this work, while recognizing the inherent importance of studying the distribution of man-made radionuclides in the marine environment, has additionally used the tracer properties of these species in the investigation of water mixing and sedimentation processes. The information thus derived has been supported in some cases by analyses of natural decay series nuclides and radiocarbon concentrations.

Previous work has not included analyses for transuranic elements and the materials studied have been confined mainly to sea water and marine sediments. Very little work has been performed on material from the extensive areas of intertidal sediments which occur in both the Clyde Sea Area and the Solway Firth, immediately to the south. These deposits can have an important function in the uptake of radioactive and stable pollutants, especially in estuarine areas. They also support a wide variety of biological species and are directly accessible for human activities, resulting in a necessity to characterize the uptake and subsequent behaviour of pollutants in this environment. A study of radionuclide

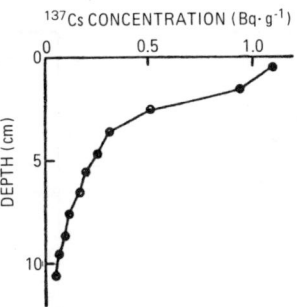

FIG.1. ^{137}Cs concentration profile for Loch Goil surface sediment, November 1977.

distributions in intertidal sediments, beach materials and coastal soils in south-
west Scotland has therefore recently been started at SURRC and the nuclides
being analysed are listed in Table I. The analytical procedure used for plutonium
is based upon that of Wong [8] and the other analytical methods used have
previously been described [9, 10].

While some intertidal sediments in this area have a relatively uniform
composition, others exhibit a highly inhomogeneous nature. For example,
particle size, chemical composition, biological activity and redox conditions can
exhibit wide variations within a small volume of sediment. These variations could
potentially have a major influence on the uptake and subsequent behaviour of
radionuclides, particularly in the case of elements such as cerium and plutonium
which can exist in a number of oxidation states which have different solubility
characteristics. Even for a nuclide such as ^{137}Cs which has a relatively simple
marine geochemical behaviour, variations in sediment properties can result in
pronounced variations in uptake. This is illustrated in Figs 1 and 2 which show
^{137}Cs concentration profiles for sediment cores collected from Loch Goil and the
Clyde Estuary in 1977 and 1978, respectively. The Loch Goil core, collected from
a site with an overlying water depth of about 90 m, showed a uniform aluminium
concentration of around 5.5% with ^{137}Cs concentrations increasing regularly
towards the surface. This increase can be related to the increase observed in the
^{137}Cs concentration of sea water in this area up to the time of collection of the
core, in response to increased discharges from Windscale. In contrast, the estuarine
sediment, collected from a site having a maximum overlying water depth of about
2 m, has an irregular ^{137}Cs concentration profile bearing no relationship to
previously observed variations in sea water concentration. In this case, the ^{137}Cs
profile does, however, show a qualitative similarity to the aluminium profile.
The latter, indicating varying clay content of the sediment, shows an irregular
concentration profile ranging from 4.2 to 5.7%. These results therefore indicate

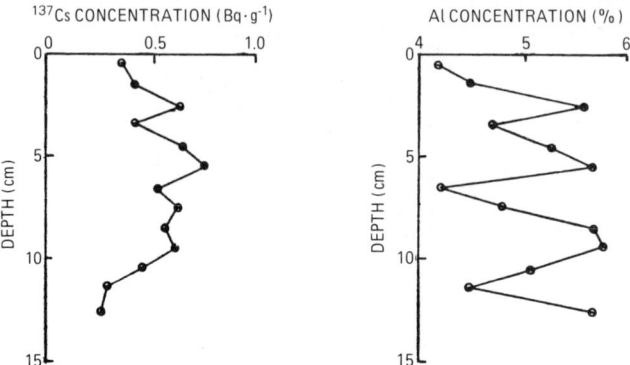

FIG.2.   [137]Cs and aluminium concentration profiles for Clyde Estuary surface sediment,
February 1978.

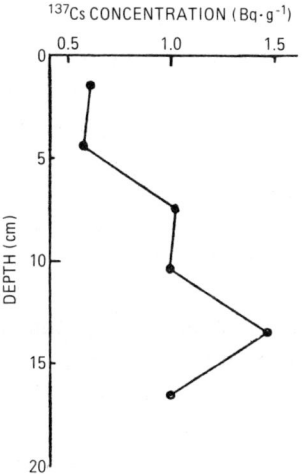

FIG.3.   [137]Cs concentration profile for Skyreburn Bay (Solway Firth) intertidal sediment,
October 1979.

the importance of chemical partitioning in determining the distribution of radionuclides in very shallow water sediments. Similar irregularities in ^{137}Cs profiles
have also been observed in some intertidal sediments from the Solway Firth as
illustrated in Fig.3.

Thus, while bulk sediment analysis is important in assessing overall
distributions of radionuclides, a more detailed knowledge of physical and chemical
partitioning within the sediment is necessary in order to understand their geochemical behaviour. Recent literature reports have indicated that very useful
information on the geochemical behaviour of plutonium in sea water can be
derived by the simultaneous use of [236]Pu and [242]Pu tracers in different oxidation

TABLE II. TYPICAL CONCENTRATIONS OF MAN-MADE
RADIONUCLIDES IN INTERTIDAL SEDIMENTS FROM
SKYREBURN BAY (SOLWAY FIRTH), OCTOBER 1979

Nuclide	Surface sediment concentration ($mBq \cdot g^{-1}$ dry sediment)
^{106}Ru	170
^{134}Cs	40
^{137}Cs	500
^{238}Pu	13
$^{239,240}Pu$	70

states [11]. Application of this method to sediment studies is, however, limited
by the chemical conditions used in the extraction of radionuclides from sediment.
Investigations of radionuclide behaviour in sediments therefore rely upon techniques
established in other areas of marine geochemistry such as the analysis of selected
physical or chemical fractions of the sediment or by comparison with the behaviour
of elements with analogous chemical properties to those of the element being
studied.

In the present work it is proposed to perform radiochemical analyses on pore
water, various particle size fractions and selected chemical leachings of the
sediment. A variety of techniques is available for this type of study as discussed
in detail elsewhere in these Proceedings. This more detailed analytical scheme
will initially be applied to intertidal sediment from.the Solway Firth, where
surface concentrations of radionuclides are sufficiently high to give detectable
levels in the various fractions of the sediment (Table II). Supporting chemical
information will be obtained using instrumental neutron activation analysis and
X-ray fluorescence analysis. These analyses will provide information on variations
in concentration of (1) major elements such as aluminium; (2) elements involved
in redox controlled solubility changes such as manganese and (3) stable counterparts of radionuclides such as caesium and cerium.

It is hoped that results obtained from this broadly based analytical approach
will provide useful information on the geochemical behaviour of radionuclides in
this important environment.

ACKNOWLEDGEMENTS

The author acknowledges the important contribution to the present work by Dr. R.D. Scott and Mrs. K.A. Harmon of SURRC and to the previous studies described by Dr. I.G. McKingley, Dr. M.S. Baxter and J. Smith-Briggs of the Chemistry Department, University of Glasgow. Dr. E.R. Sholkovitz is thanked for helpful discussions.

REFERENCES

[1] NATURAL ENVIRONMENT RESEARCH COUNCIL, The Clyde Estuary and Firth. An Assessment of Present Knowledge Compiled by Members of the Clyde Study Group, NERC Publications Series C, No. 11 (1974).

[2] CLYDE RIVER PURIFICATION BOARD, Annual Reports, Rivers House, East Kilbride, Glasgow.

[3] BAXTER, M.S., HARKNESS, D.D., "$^{14}C/^{12}C$ ratios as tracers of urban pollution", Isotope Ratios as Pollutant Source and Behaviour Indicators (Proc. IAEA/FAO Symp. Vienna, 1974), IAEA, Vienna (1975) 135.

[4] CAMBRAY, R.S., JEFFERIES, D.F., TOPPING, G., An Estimate of the Input of Atmospheric Trace Elements into the North Sea and the Clyde Sea (1972–3), UKAEA Publication AERE-R7733 (1975).

[5] BAXTER, M.S., McKINLEY, I.G., "Radioactive species in sea water", Proc. R. Soc. Edinb. **768** (1978) 17.

[6] BAXTER, M.S., McKINLEY, I.G., MacKENZIE, A.B., JACK, W., Windscale radiocaesium in the Clyde Sea Area, Mar. Pollut. Bull. **10** (1979) 116.

[7] SMITH-BRIGGS, J.L., Pollution studies in the Clyde Sea Area, Anal. Proc. **17** (1980) 5.

[8] WONG, K.M., Radiochemical determination of plutonium in sea water, sediments and marine organisms, Anal. Chim. Acta **56** (1971) 355.

[9] MacKENZIE, A.B., BAXTER, M.S., McKINLEY, I.G., SWAN, D.S., JACK, W., The determination of ^{134}Cs, ^{137}Cs, ^{210}Pb, ^{226}Ra and ^{228}Ra concentrations in nearshore marine sediments and seawater, J. Radioanal. Chem. **48** (1979) 29.

[10] MacKENZIE, A.B., SCOTT, R.D., Separation of bismuth-210 and polonium-210 from aqueous solutions by spontaneous adsorption on copper foils, Analyst **104** (1979) 1151.

[11] NELSON, D.M., LOVETT, M.B., Oxidation state of plutonium in the Irish Sea, Nature **276** (1978) 599.

USE OF RESIDENCE TIME MODELS
IN ECOLOGICAL STUDIES OF TRANSURANICS

M.J. FRISSEL, F. van DORP, P. POELSTRA
Association EURATOM-ITAL,
Wageningen,
Netherlands

Abstract

USE OF RESIDENCE TIME MODELS IN ECOLOGICAL STUDIES OF TRANSURANICS.
   The paper discusses the applicability of different types of ecological models. An
explanation of the various concepts and the basis for residence time models and how they are
correlated is presented. *Mean residence time* and *pseudo residence time* are terms useful for
more complicated systems. The term *chromatography* defines a conceptual model, but the
processes involved are limited to diffusion, dispersion, convection, adsorption and desorption.
The most sophisticated type of models are the simulation models. Regression equations are
used to calculate the transfer between pools of simulation models. However, simulation models
consider processes in great detail and processes are described by appropriate equations. Lack
of useful parameter values is the main limitation for applying such models. Their main value
is in promoting understanding of radioecological processes and in identifying missing information.

1.   INTRODUCTION

One of the merits of radioecology is that it has initiated investigations into
harmful effects of radionuclides on the ecosystem before any recognizable damage
had occurred. In the absence of any observed damage one can only speculate on
the most valuable species of the ecosystem. A start can be made to provide a
quantitative description of pathways and transfers through the different
components of the environment by building suitable models.

Because of their simplicity, residence time models are rather popular among
radioecologists. The term *residence time* has a distinct physical meaning and
should, therefore, not lead to any misinterpretation. A disadvantage is that the
use of residence times does not provide understanding of the underlying processes,
which are assumed to control the residence time. Misunderstanding may occur
because some users refer (with further indication) to specific first-order rate kinetics
models, others to specific steady-state models.

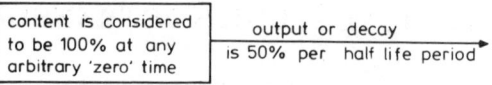

FIG.1.   A half-life time model.

2.   THE RESIDENCE TIME IN 'HALF-LIFE TIME' MODELS

The classic example of a half-life time model (Fig.1) is an amount of a radio-nuclide which disintegrates because of its radioactive decay. The decay rate at any instant is proportional to the quantity present

$$dN/dt = -\lambda N \tag{1}$$

where N is the amount of radioactivity and λ the decay constant. Integration leads to the definition of

$$T_{\frac{1}{2}} = 0.693/\lambda \tag{2}$$

where $T_{1/2}$ is the period (in units of time) in which 50% of the pool content decays.

The same approach can be applied to other biological, chemical or physical processes, provided their overall and effective disappearance rates are controlled by first-order rate kinetics. A classic example is the biological excretion of radio-active material, where both half-life times can be coupled by

$$T_{eff} = T_{1/2} \cdot T_b/(T_{1/2} + T_b) \tag{3}$$

where T_{eff} is the effective disappearance half-life time and T_b the excretion (or biological) half-life time. All values are in units of time.

The mean period that some compound remains in the pool or its average life-time is given by

$$\tau = T_{1/2}/0.693 = \frac{1}{\lambda} \tag{4}$$

τ is called residence time, which is misleading because in fact most of the compound in the pool remains for a considerably longer or shorter period within the pool. Winteringham [1] uses therefore the term *mean residence time*.

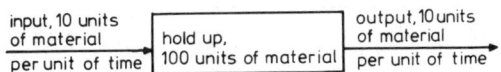

FIG.2.   *A steady-state model.*

3.   THE RESIDENCE TIME IN STEADY-STATE MODELS (TURNOVER TIME MODELS)

A classical example of a steady-state model (Fig.2) is a lake containing a constant amount of soluble radioactivity. The flux of radioactivity both into and out of the lake is equal and constant with time. The quantity in the lake or, more generally, in the pool, is defined as the holdup. Because input and output are equal, it is not necessary to distinguish between them, thus the term through-put can be used. The residence time τ is defined as

$$\tau \text{ (units of time)} = \frac{\text{holdup (units of radioactivity in the pool)}}{\text{throughput (units of radioactivity per unit of time)}} \quad (5)$$

τ is well-defined only for steady-state conditions, the nature of the input and output processes is not considered. In case of piston-flow within the pool, the period that all material stays within the pool equals the residence time. When mixing occurs, the actual residence period may be somewhat shorter or longer than the *residence time;* nevertheless, the residence time is a very realistic one. Combinations of steady-state pools are described by Gibilaro [2]. Both a sequence of pools or parallel pools can be included.

4.   COMBINATION OF HALF-LIFE TIME MODEL AND STEADY-STATE MODEL

A combination of a steady-state throughput process and a radioactive decay process may occur (Fig.3). The residence time of the material within the pool is according to Eq.(5) 5 units of time. If the radioactive decay is considered separately, according to Eq.(1), a λ of 0.02 (time^{-1}) can be calculated. $T_{1/2}$ is, accordingly, 34.65 units of time and the residence time 50 units of time (Eq.(4)). From a physical point of view this is acceptable because the calculation of τ for the radioactive material by Eq.(5), only considering the pool holdup of 100 and the output by decay of 2, gives the same result, namely a residence time of

FIG.3.   A combination of steady state and radioactive decay.

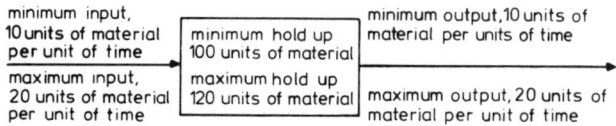

FIG.4.   A steady-state model with fluctuating inputs, outputs, and holdup.

50 units of time. This example shows that an output by first-order rate reactions does not provide problems as long as steady conditions are maintained.

Turnover time. Another expression for the residence time as defined in the Figs 2 and 3 is turnover time. In a classic paper on turnover functions of labelled agents Zilversmit et al. [3] defined turnover time as the time that 'a substance in a tissue required for the appearance or disappearance of an amount of that substance equal to the amount of that substance present in the tissue'. This definition is equivalent to Eq.(5).

5.   MEAN RESIDENCE TIME

In ecological and biological models the requirement of steady-state conditions is almost never fulfilled. Sometimes, however, inputs, outputs and holdups fluctuate between certain levels. In the example which may represent a lake with inputs and outputs of radiocontaminants via a river, the residence time for minimum conditions is $100/8 = 12.5$ units of time and for maximum conditions is $120/25 = 9$ units of time (Fig.4). When the periods to which minimum and maximum conditions apply last t_n and t_m, respectively, it seems reasonable to define a mean residence time $\bar{\tau}$ as

$$\bar{\tau} = \frac{t_n}{t_n + t_m} \left(\frac{\text{minimum holdup}}{\text{minimum throughput}} \right) + \frac{t_m}{t_n + t_m} \left(\frac{\text{maximum holdup}}{\text{maximum throughput}} \right)$$

$$(6)$$

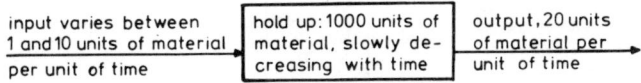

FIG.5. *Pseudo steady-state model. The stability of the system depends on the large pool of which only a small fraction is used/released per unit of time.*

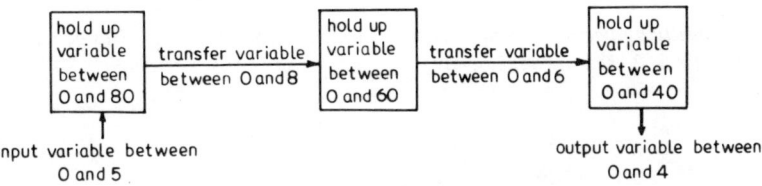

FIG.6. *Pseudo steady-state model. A sequence of pools. The transferred amounts are a function of the holdup in the pool which results in a certain degree of stability.*

6. PSEUDO RESIDENCE TIME

Most biological or ecological systems are neither steady-state systems nor first-order decay rate systems, but rather slowly changing dynamic systems. A first example may be organic material in a lake sediment, which owing to changing conditions slowly decomposes. In this case the organic material is a comparatively large pool with only small inputs and somewhat larger outputs (Fig.5).

Provided the organic matter pool is large enough, a residence time can be defined in a similar way as for a half-life time model. Because of poorly defined conditions *pseudo residence time* may be a more appropriate term than residence time itself.

A second example refers to a radiocontaminant which passes a sequence of lakes and rivers (Fig.6). Inputs are almost never a function of the holdup. The outputs depend on the holdup, which stabilizes the system. These slowly changing systems are often the main interest of ecologists and biologists. Also for these systems a pseudo residence time τ' can be defined.

$$\tau' = \frac{\text{holdup (units of material within the pool)}}{\text{output (units of material per unit of time)}} \tag{7}$$

The definition has a physical meaning only when the output per unit of time is small compared to the holdup. In this case neither the input nor the throughput

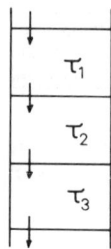

FIG.7.   A multicompartment residence time model. $\tau_1, \tau_2, \tau_3, ... \tau_n$ = residence time in the compartments 1, 2, 3, ... n, respectively.

can be used to define τ' in a useful way. An advantage of the term pseudo residence time is that its physical meaning is equal to the residence time as defined by Eqs (4) or (5). It makes a comparison of different types of ecological or biological systems easier.

7.   MULTICOMPARTMENT MODELS AND CHROMATOGRAPHY MODELS

A special form of pseudo residence times may be used to describe the migration of radiocontaminants through sediments or soils. Usually only one direction (the vertical one) is taken into account.

An example of such a multilayer pseudo residence time model is shown in Fig.7. It is very typical that the mechanisms on processes which control the residence time are not considered.

In contrast to multicompartment residence time models, chromatography models do consider the processes explicitly. A simplified definition of chromatography may be the following:

Chromatography is a combination of transport, adsorption and desorption processes in a system consisting of a moving liquid phase and a non-moving solid phase. The actual transport of the compounds under consideration occurs within the liquid phase; because of adsorption the transport rate of the compounds under consideration is retarded compared to the transport rate of the liquid phase. Different compounds may have different retardations. Chromatography is often described by a one-dimensional multicompartment model in which massflow and diffusion flow can be distinguished (Fig.8).

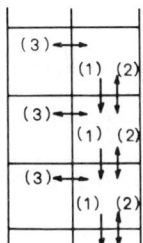

FIG.8. *Chromatography. The liquid phase is assumed to move downwards. (1) = transport by massflow; (2) = transport by diffusion (bidirectional); (3) = adsorption and desorption.*

Almost without exception chromatography is described by the Focke-Planck equation or one of its modifications:

$$\frac{\delta C_s}{\delta t} + \frac{\delta C_a}{\delta t} = D \frac{\delta^2 C_s}{\delta x^2} - v \frac{\delta C_s}{\delta x} \tag{8}$$

where t = time
      x = distance
      C_s = concentration of compound in solution phase (mol/cm^3)
      C_a = concentration of compound in adsorber phase (mol/g)
      D = apparent diffusion constant of compound in solution phase
      v = linear velocity of solution in x-direction.

The transfer between the different layers occurs by massflow and diffusion. The diffusion into a certain layer n depends on the concentration gradient between layer n and layer n-1. One of the conditions of a pseudo residence time is that the output from layer n should not be a function of the concentration within layer n-1. This condition is not fulfilled in a chromatography model and the latter model differs therefore from a pseudo residence time model. Often the diffusion is negligible compared to the massflow; in that case the chromatography model comes very close to a multilayer pseudo residence time model. To calculate the time of residence in a certain layer the following equation is proposed:

$$\tau' = \frac{1}{v} + \frac{1}{v} \cdot \frac{C_a}{C_s} \cdot \frac{\rho}{\theta} \tag{9}$$

where ρ is the density in g per cm^3 (dry) bulk soil and θ is the moisture content in ml water per cm^3 (wet) bulk soil.

In northern american soil science literature chromatography is often called miscible displacement.

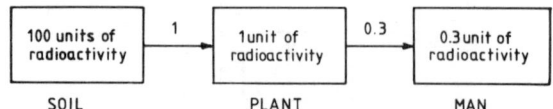

FIG. 9.   A very simple relational diagram.   Only some transfers are indicated, other transfers are neglected.

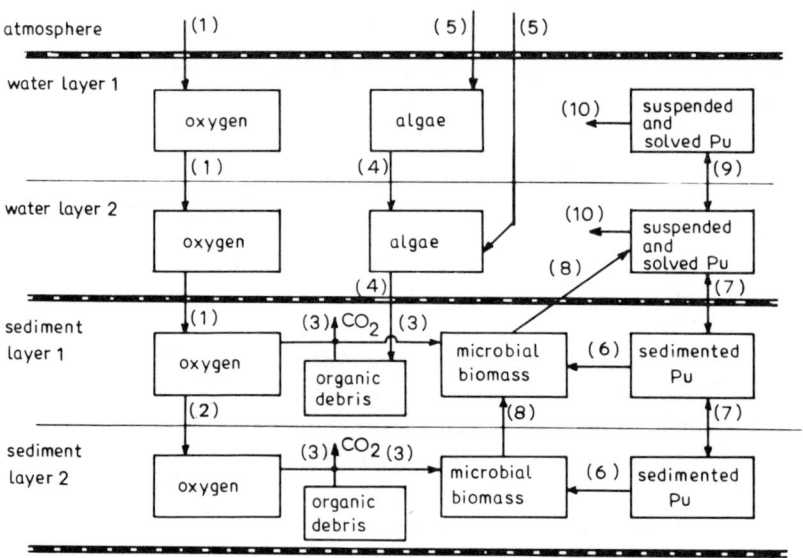

FIG. 10.   Model to describe the influence of oxygen on the uptake of plutonium.   For simplicity, the water and sediment layer are divided into two layers, for computations they should be divided into more layers.   Atmospheric oxygen is by diffusion and mixing transferred from the atmosphere to the sediment layer (1); within the sediment oxygen is transferred by diffusion only (2).   This oxygen is within the sediment layer used as electron acceptor by the growing microorganisms; organic debris products serve as carbon substrate (3).   For the upper layer organic debris stems from decaying algae (4); the growth of algae is determined by solar radiation (5).   Within the second sediment layer organic debris is slowly depleted. When the oxygen consumption within a particular sediment layer is higher than the supply of oxygen, anaerobic conditions prevail.   This induces a lower redox potential which may lead to solution or resuspension of sedimented plutonium (7).   The growing microbial biomass incorporates sedimented plutonium (6), which via decomposition of the microorganisms reaches the water phase as organically bound plutonium (8).   Via mixing and diffusion resuspended plutonium reaches the upper water layers (9) where plutonium may be taken up (10).

8.   VERY SIMPLE RELATIONAL DIAGRAM

Figure 9 shows on oversimplified relational diagram; it is often used to calculate contaminations of the food chain. Only one transfer reaction is considered, the other ones are neglected. It is impossible to calculate from such a transfer factor a residence time; the residence time concept is simply not applicable. From an ecological point of view these types of models are very poor and discussions and sophisticated statistical studies on the real values of the transfer factors are therefore rather exaggerated.

9.   SIMULATION MODELS

The power of all models described so far is rather low. They do not allow handling of complex systems. For a proper description of complex systems simulation models are the most suitable ones. Figure 10 describes as an example a system which takes into account interactions between solar radiation, growth of microorganisms, oxygen diffusion and dispersion through surface water, anaerobic conditions induced by depletion of the oxygen pool, changes of the concentration of transuranics in the waterphase due to anaerobic conditions, migration and uptake of transuranics. Many pools are involved and most of the transfers between the pools have a dynamic character. It is very remarkable that such a model immediately provokes (negative as well as positive) criticism, while a much simpler model as shown in Fig.9 does not. That is understandable, the model as shown in Fig.10 is a real concept, it stimulates an in-depth analysis of the system modelled. Indeed, the model is more a tool to promote understanding and to identify missing information than a forecasting model. To describe all transfers between the pools of Fig.10 mathematically, two types of approach are commonly used.

The first type of approach applies regression equations in which the transfers are related to pool sizes and auxiliary conditions (temperature, pH value, redox potential, nutrient status etc.), without considering the processes which are responsible for the transfers. The parameters for the regression equations are often derived by (very accurate) statistical methods from (poorly reviewed) literature reports. The approach has particular advantages when forecasting is the main objective of the model. The main disadvantage of regression models is that it is often not possible to apply coefficients which were derived for a number of areas to other areas because these other areas may be different from a geographical or climatological point of view.

The second type of approach concentrates on the underlying concept of the model. It attempts to understand all processes and mechanisms in detail and to describe them by equations which are applicable to such processes. For each process description there are usually many options depending on the degree of sophistication one whishes to apply. Growth of microorganisms can be considered as a steady-state process, as a first-order kinetics process or as a Michaelis-Menten type process. The growth rate can be modified as function of substrate availability (carbon, nitrogen, phosphates), temperature and other environmental conditions. Again, depending on the degree of sophistication, a simple death rate of micro-organisms can be assumed or a refined maintenance energy model can be developed. For all other relations the same options exist. For example it is important to note that both reversible and irreversible plutonium transformations can be handled conveniently. In the first case the status of plutonium is directly controlled by the environmental conditions (pH value, redox potential, concen-tration of complexing agents); in the second case rate parameters control the transfer from one status to the other one.

A serious drawback of this type of approach is that it is sometimes very time-consuming to collect the data which are required. It must be expected that this will only slightly improve in the next five or ten years so that these models despite their high degree of sophistication cannot yet be applied for forecasting. Their main value lies, for the time being, in promoting understanding of (radio-)ecological processes.

REFERENCES

[1]   WINTERINGHAM, F.P.W., Comparative ecotoxicology of halogenated hydrocarbon residues, Ecotoxicology and Environmental Safety 1 (1977) 407.
[2]   GIBILARO, L.G., Mean residence times in continuous flow systems, Nature 270 (1977) 47.
[3]   ZILVERSMIT, D.B., ENTEMON, C., FISHLER, M.C., On the calculation of 'turnover time' and 'turnover rate' from experiments involving the use of labelling agents, J. Gen. Physiol. 26 (1942) 325.

INTERCOMPARISON OF
TRANSURANICS IN SEDIMENTS*

K. NILSSON
Risø National Laboratory,
Roskilde,
Denmark

Abstract

INTERCOMPARISON OF TRANSURANICS IN SEDIMENTS.
This paper summarizes the results of a recent intercomparison of transuranic analyses in Nordic sediments undertaken by five laboratories. Five different samplers were tested to determine their consistency and utility. Plutonium and americium were determined in all the samples and the analysis of variance is discussed.

In May 1979, five Nordic laboratories undertook an intercomparison experiment on sampling and analysis of sediments. The samples were collected in Øresund at a position about 4 km north-west of Barsebäck. Øresund is the sound between Sweden and Denmark.

Representatives from a Finnish, a Norwegian, two Swedish and a Danish laboratory participated, and each laboratory supplied and handled its own sampler. Two of these were identical, but otherwise there was considerable variation in the general design. The sample collectors were cylindrical in shape with diameters ranging from 3.3 to 13.6 cm, and all but one could collect samples to a depth of 10 cm. In order to obtain sufficient material, it was necessary to collect 30 cores with the smallest sampler, whereas 4 cores were enough with the largest sampler. All cores were immediately divided into three sections, 0–2 cm, 2–4 cm and 4–10 cm (or 4–6 cm). The material from identical sections of a sampler was pooled, then dried and weighed.

The relevant data concerning the material is presented in Table I. The weight per square unit for the same sample sections from the different collectors is similar, although the two samplers of identical design seem to collect somewhat less material per square unit than the other samplers. The reason for this minor difference is unclear. The total surface area collected by the different samplers varied by a factor of two, but it is to be noted that the smallest and the largest sampler collects the same amount of material per square unit.

---

* The study was supported by the Nordic Liaison Committee for Atomic Energy.

TABLE I. SAMPLE MATERIAL

Sampler	Sample collector ϕ (cm)	Sample section (cm)	Number of cores	Total area (cm^2)	Total dry weight (g)	Weight/area (g·cm^{-2})
		0–2			192.2	0.814
S$_1$	5	2–4	12	236	225.6	0.956
		4–10			769	3.258
		0–2			578.6	0.998
S$_2$	13.6	2–4	4	580	645.6	1.113
		4–10			2045.0	3.526
		0–2			166.0	0.703
S$_3$	5	2–4	12	236	212.4	0.900
		4–10			745.7	3.160
		0–2			472.3	0.948
S$_4$	6.5	2–4	15	498	573.2	1.151
		(4–6)			616.1	(1.237)
		0–2			248.4	0.967
S$_5$	3.3	2–4	30	257	308.1	1.199
		4–10			904.5	3.519

Ten gram aliquots of each section of the sediments were distributed to the five laboratories and analysed for ^{239,240}Pu; in two of the laboratories also for ^{241}Am.

The remaining material was circulated among the laboratories for measurement of ^{137}Cs, ^{60}Co and ^{40}K by Ge(Li) spectrometry.

In general, the laboratories followed the well-established method of Talvitie [1] in the radiochemical analysis for ^{239,240}Pu. ^{241}Am was determined by a method described by Holm et al. [2].

The content of ^{239,240}Pu in the sediments is presented in Table II; the content of ^{241}Am, which was measured in two of the laboratories, is presented in Table III. The data are given in Bq·kg^{-1}.

TABLE II. ^{239,240}Pu IN SEDIMENTS COLLECTED WITH FIVE DIFFERENT
SAMPLERS AND ANALYSED IN FIVE DIFFERENT LABORATORIES (Bq·kg^{-1})

Depth	Sampler	Analytical laboratory				
		A 1	A 2	A 3	A 4	A 5
	S 1	2.52	2.52	3.03	2.22	–
	S 2	3.44	2.18	2.85	2.07	2.92
0–2 cm	S 3	1.52	1.70	3.03	2.04	2.15
	S 4	1.92	2.33	2.89	2.55	2.92
	S 5	2.85	2.37	3.03	2.26	3.00
	S 1	1.96	1.92	2.81	1.81	2.26
	S 2	3.29	1.73	2.81	1.59	–
2–4 cm	S 3	2.26	2.26	2.85	2.00	2.04
	S 4	1.78	1.48	2.15	2.00	2.81
	S 5	3.15	2.41	3.44	2.70	2.96
	S 1	–	–	0.15	0.15	0.33
	S 2	–	0.11	0.44	0.15	0.41
4–10 cm	S 3	–	0.22	0.52	0.37	0.85
(4–6) cm	S 4	–	(0.48)	(0.67)	(0.52)	(0.89)
	S 5	–	0.44	0.56	0.52	0.93

There is little difference in the average ^{239,240}Pu content of the two upper
layers. The 0–2 cm section contains 2.51 ± 0.49 Bq·kg^{-1}, and the 2–4 cm
section contains 2.35 ± 0.56 Bq·kg^{-1}. The lowest section, 4–10 cm, has a
^{239,240}Pu content of 0.41 ± 0.25 Bq·kg^{-1}.

The average values for the ^{241}Am content of the three sections of the
sediment are 0.62 ± 0.12 Bq·kg^{-1} for the 0–2 cm layer, 0.65 ± 0.13 Bq·kg^{-1}
for the 2–4 cm layer and 0.13 ± 0.08 Bq·kg^{-1} for the 4–10 cm layer.

In the bottom layer the limit of detection in 10 g aliquots of sediment is
approached for ^{239,240}Pu as well as for ^{241}Am, which is reflected in a greater
increased relative standard deviation for the values of the 4–10 cm section.

TABLE III. ^{241}Am IN SEDIMENTS COLLECTED WITH FIVE DIFFERENT
SAMPLERS AND ANALYSED IN TWO DIFFERENT LABORATORIES (Bq·kg^{-1})

Depth	Sampler	Analytical laboratory	
		A 2	A 3
	S 1	0.59	0.70
	S 2	0.78	0.59
0–2 cm	S 3	0.56	0.81
	S 4	0.52	–
	S 5	0.48	0.52
	S 1	0.52	0.67
	S 2	–	0.48
2–4 cm	S 3	0.81	–
	S 4	–	–
	S 5	0.63	0.78
	S 1	0.04	0.11
	S 2	0.22	0.07
4–10 cm	S 3	0.07	0.11
(4–6) cm	S 4	0.11	–
	S 5	0.11	0.30

The ^{241}Am/^{239,240}Pu ratio for the three sections of the sediments amounts
to 0.25 for the 0–2 cm layer, to 0.28 for the 2–4 cm layer and to 0.32 for
the 4–10 cm layer. The ^{241}Am/^{239,240}Pu ratio in the global fallout inventory
has been calculated to have a current value of 0.2 to 0.3. The findings of
this investigation agree with these values and it can be concluded, as expected,
that the two actinides, ^{239,240}Pu and ^{241}Am determined in these sediments from
Øresund stem from global fallout.

The plutonium values were subjected to analysis of variance; the result is
presented in Table IV. The analysis of variance on all layers shows a highly
significant difference between the ^{239,240}Pu content of the three sections of
the sediments, but there is no difference between the results from the five
analytical laboratories nor between the contents of plutonium found when
different samplers were used. The analysis of variance indicates no second-order
interaction between the three variables: depth, sampler and analytical laboratory.

TABLE IV. ANALYSIS OF VARIANCE OF ^{239,240}Pu (Bq·kg^{-1})

Analysis of variance on all sample sections

	Depth	* * *
Main effects	Sampler	–
	Analysis	–
	D × S	* * *
	D × A	* * *
Interactions	A × S	*
	D × A × S	–

Analysis of variance on each sample section

		0–2 cm	2–4 cm	4–10 cm
	Sampler	–	*	* * *
Main effects	Analysis	*	*	* *
Interactions	A × S	–	* *	–

*: Probably significant (P ⩾ 95%); **: Significant (P ⩾ 99%); ***: Highly significant
(P ⩾ 99.9%).

There is, however, a significant first-order interaction between depth and the
samplers, and between depth and the analytical laboratories.

The analysis of variance on each of the sample layers shows that the
^{239,240}Pu content in the bottom layer differs significantly from the upper
layers both as a function of the analytical work and as a function of the different
samplers. This can be seen in Figs 1 and 2.

Figure 1 presents for each layer the average value for the five samplers as
determined in the five analytical laboratories. The average values have been
normalized to 1 for each layer. The values for the two upper layers cluster
around the average of 1, although there is a tendency for the same laboratory
to determine systematically either slightly higher or slightly lower values than
the average. The difference among the analytical laboratories is apparent in
the results from the bottom layer. The explanation is probably to be found in
the ^{239,240}Pu content being about five times lower than in the upper layers.
The limits of detection for ^{239,240}Pu in 10 g of sediment is then approached.

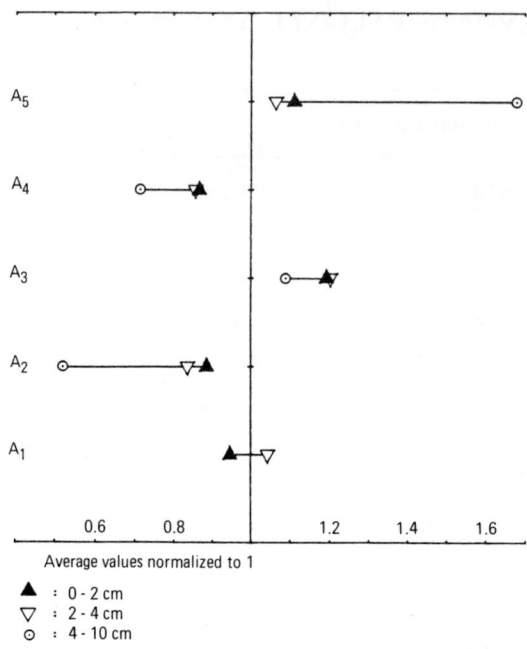

FIG.1. ^{239,240}Pu content in sediments as a function of the analytical laboratory.

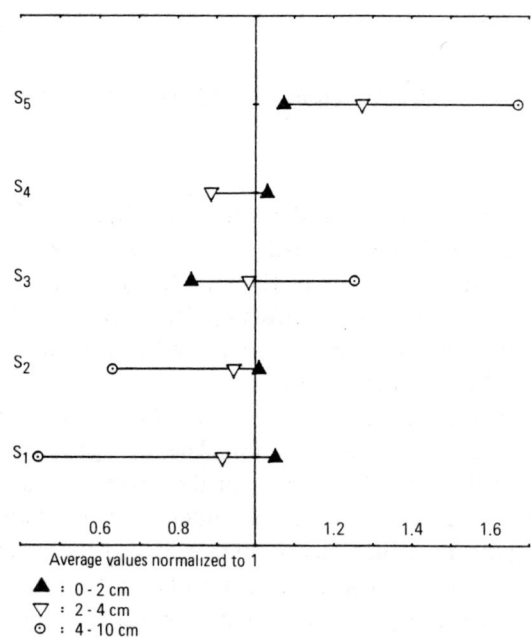

FIG.2. ^{239,240}Pu content in sediments as a function of the different samplers.

The analysis of variance on each of the sample sections also shows a significant difference in the ^{239,240}Pu content between the upper layers and the bottom layer among the five samplers as determined by the different laboratories. This can be seen in Fig.2. The values for the upper two layers cluster at random around the average value of 1, whereas there is much more spreading among the values for the bottom layer.

The result with sampler 5 differs slightly from the results with the other samplers, since there seems to be a relative increase in the content of ^{239,240}Pu with depth. The explanation probably is that the possibility of the lower layers being slightly contaminated with sediment from the upper layers, which have a five times higher actinide concentration, is increased with the number of cores being taken. Sampler 5 collected by far the largest number of cores during the experiment. Sampler 1 and sampler 3 are identical and a comparison of these two samplers indicates the errors which stem from handling the equipment and analysing the samples. Such variations are to be expected from an intercomparison experiment with all the uncertainties involved.

It is surprising and stimulating that an investigation, which involves both collection of samples and analytical work to determine tracer amounts of material, all carried out by different people, can yield results which agree so well.

REFERENCES

[1] TALVITIE, N.A., Anal. Chem. **43** (1971) 1827.
[2] BALLESTRA, S., HOLM, E., FUKAI, R., Determination of Radionuclides in Environmental and Biological Materials (Proc. Symp. Central Electricity Generating Board, October 1978, London, paper No.15).

SUMMARY AND RECOMMENDATIONS

1. GENERAL DISCUSSION

Papers presented at the meeting in this section ranged from direct discussions of the interaction of transuranic elements with sedimentary systems to the more general considerations of the partitioning of trace elements in sediments. A theme common to almost all papers, however, was the exchangeability of transuranic elements between the aqueous phase and the sediment. The discussion group recognized the major role of sediments in all aquatic environments for binding transuranic elements, whatever their origin, and there was agreement that laboratory studies, to investigate the extent and rate of release of the elements from sediments to associated waters, should be continued along the lines suggested in several papers at the meeting. It was also felt that leaching experiments should be extended to cover other types of environment not yet being studied.

With due regard to the importance of the oxidation state of transuranic elements in relation to their distribution between aqueous and solid phases, the group recognized the need to determine the oxidation/reduction (or redox) capacity of an environment in order to understand and predict the behaviour of these elements. These measurements are equally applicable to interactions at the sediment/water boundary and to the post-depositional behaviour of the elements after burial. In this context the group wished especially to encourage the further development and comparison of techniques for the study of interstitial waters. It became clear during discussions that the study of ground water/rock interactions are vital to a proper assessment of the fate of both transuranic and other radionuclides arising from disposal sites on land; these also require a knowledge of redox conditions as well as other parameters.

The role of organic ligands as solubilizing agents for transuranic elements was highlighted both in the papers presented and in round-table discussions. The need for further work on this topic was noted and the group agreed that information was required from locations other than those which had already been studied.

There was some disagreement as to the applicability of chemical leaching techniques to study the partition of transuranics between the various phases of a sediment. Unlike stable elements where there is a need to distinguish between the residual and non-residual fractions in order to assess the pollutant load, the origin of the man-made elements is undisputed. Nevertheless, it was considered useful to carry out partitioning studies in order to determine the main phases with which transuranic elements associate in different situations. Differing opinions were expressed about the use of normalizing elements (such as aluminium or scandium) to assist in comparing the concentration of elements found on sediments of varying types. There was general agreement that such techniques can be helpful, particularly as complementary procedures with the physical separation of sediments into size fractions, to facilitate the comparison of radionuclide levels in different types of sediment.

Considerable attention was paid to the importance of relating knowledge of oceanic sediments occurring in the vicinity of deep sea dumping sites (much of which already exists) to the likely behaviour of transuranic and other radionuclides released onto sediment surfaces from canisters, or released from within the sediment following the emplacement of high-level waste in bore holes. Apart from the general assessment of sediment types and the physico-chemical conditions which prevail at any site, the group considered the possible use of analogue elements such as the lanthanides uranium and thorium to help predict the behaviour of transuranic elements. It was generally felt that great care was needed in the use of such analogues due to oxidation state changes and the different degrees of complexation of the various elements with both organic and inorganic ligands. It was realized therefore that all possible use should be made of the known behaviour of transuranics and any suitable analogues in the coastal and other sites where they can all be studied together in relation to the prevailing environmental conditions. The comparative data so obtained can then at least be of some use in improving our understanding of transuranic behaviour in the deep water environments.

The role of macro- and micro-biological activity in determining the chemical forms and migration of transuranics was considered to be inadequately understood. It was nevertheless realized that bioturbation can play an important role in the translocation of transuranics in the surface layers of sediments. The need for greater co-operation between biologists and sedimentologists was seen as the key to improving knowledge in this field. Reference was made to the use of X-ray analysis of cores and the identification of burrowing organisms as well as work on bacterial activity in sediments, all of which it was felt could be used to enhance the present work.

When considering the fate of transuranic nuclides in the environment a distinction must be made between their short-term behaviour and their ultimate fate in the long term (say $\geqslant 100$ years). In dealing with time-scales such as these, the modelling of those processes which control the ultimate fate of transuranics in sedimentary as well as in aquatic environments is an essential topic. Severe limitations are imposed on such modelling at the present time due to an inadequate data base. It was thus realized that the acquisition of suitable data is now of paramount importance if our understanding of the long-term impact of the man-made elements on the environment is to be predicted with any degree of certainty.

2. RECOMMENDATIONS

(1) The extent and rate of release of transuranic elements from sediments to associated waters should be studied for a variety of sites.

(2) The redox capacity of the various environments needs to be investigated so that the effect of oxidation state changes on the distribution of transuranics

between sediment and water phases can be adequately assessed. Ground waters should be studied for land disposal sites.

(3)   Comparative geochemical studies should be made of deep-water dumping sites and those coastal water sites which now receive transuranic wastes so that the possible behaviour of these elements in oceanic environments can be assessed.

(4)   The role of organic ligands as solubilizing agents, and their effect on post-depositional mobility needs further investigation. Their effect on bioavailability of transuranics is also important.

(5)   The effect of bioturbation in the translocation of transuranic and other radionuclides in surface sediments needs further work, as does the role of bacteria on the fate of these elements in sediments.

(6)   Data suitable for the construction of models to predict the long-term behaviour of transuranics in the environment needs to be collected. The importance of co-operation between those producing data and the modelling user needs to be emphasized.

CHAIRMEN OF SESSIONS

Session I   D.N. EDGINGTON   United States of America
                 E. HOLM   Sweden

Session II   O. VANDERBORGHT   Belgium
                 J. PENTREATH   United Kingdom

Session III   R. CHESTER   United Kingdom
                 E. DUURSMA   Netherlands

SECRETARIAT

Scientific   W. FORSTER*   Division of Waste Management,
  Secretaries: (Session I)   IAEA, Vienna
                 C. MYTTENAERE
                 (Session II)   CEC, Brussels
                 C.N. MURRAY   Health Physics Division,
                 (Session III)   CEC, Ispra

Editor: B. KAUFMANN   Division of Publications,
                                           IAEA, Vienna

---

* *Present address:* ER–75, Department of Energy, Washington, DC 20545, USA.

LIST OF PARTICIPANTS

**METHODS OF DETERMINING TRANSURANIC CHEMICAL SPECIATION
AT ENVIRONMENTAL LEVELS $\leqslant 10^{-10}$ M**
(Session I)

Avogadro, A.*	CEC, Joint Research Centre, Chemistry Division, I-21020 Ispra (Varese), Italy
Ballestra, S.*	International Laboratory of Marine Radioactivity, Oceanographic Museum, Monaco-Ville, Principality of Monaco
Bidoglio, G.	CEC, Joint Research Centre, Chemistry Division, I-21020 Ispra (Varese), Italy
Billon, A.*	CEA, Centre d'études nucléaires de Fontenay-aux-Roses, Service d'études analytiques, B.P. 6, F-92260 Fontenay-aux-Roses, France
Delle Site, A.*	CNEN, CSN Casaccia, CP 2400, I-00100 Rome, Italy
Edgington, D.N.*	Centre for Great Lake Studies, University of Wisconsin-Milwaukee, P.O. Box 413, Milwaukee, WI 53201, United States of America
Eicke, H.-F.*	Deutsches Hydrographisches Institut, Wüstland 2, D-2000 Hamburg 55, Federal Republic of Germany
Forster, W.	International Atomic Energy Agency, P.O. Box 100, A-1400 Vienna, Austria
Fukai, R.*	International Laboratory of Marine Radioactivity, Oceanographic Museum, Monaco-Ville, Principality of Monaco
Holm, E.*	Radiation Physics Department, University of Lund, Lasarettet, S-221 85 Lund, Sweden

---

* Presented Paper in Plenary Session.

Lovett, M.B.*

Ministry of Agriculture, Fisheries and Food,
Directorate of Fisheries Research,
Fisheries Radiobiological Laboratory,
Hamilton Dock,
Lowestoft, Suffolk NR32 1DA,
United Kingdom

Martin, J.M.*

Laboratoire de géologie, Ecole normale supérieure,
46 rue d'Ulm, F-75230 Paris Cedex 05, France

Nilsson, K.*

Risø National Laboratory,
DK-4000 Roskilde, Denmark

Piro, A.

CNEN, Laboratorio Studio Ambiente Marino,
I-19030 Fiascherino- La Spezia, Italy

Schweingruber, M.

EIR (Federal Institute for Reactor Research),
CH-5303 Würenlingen, Switzerland

METHODS OF STUDYING THE BIOAVAILABILITY OF TRANSURANICS IN AQUATIC SYSTEMS
(Session II)

Angeletti, L.*

CEA, Centre d'études nucléaires de Fontenay-aux-Roses, DPr
B.P. 6, F-92260 Fontenay-aux-Roses, France

Hoppenheit, M.

Biologische Anstalt Helgoland,
Wüstland 2, D-2000 Hamburg 55,
Federal Republic of Germany

Metayer-Piret, M.*

CEN/SCK,
Boeretang 200, B-2400 Mol, Belgium

Myttenaere, C.

CCE, Biologie, Radioprotection,
rue de la Loi 200, B-1040 Brussels, Belgium

Pentreath, R.J.*

Ministry of Agriculture, Fisheries and Food,
Directorate of Fisheries Research,
Fisheries Radiobiological Laboratory,
Hamilton Dock,
Lowestoft, Suffolk NR32 1DA,
United Kingdom

Pieri, J.

Faculté des Sciences, Université de Nantes,
Laboratoire de biochimie,
2, Chemin de la Houssinière,
F-44000 Nantes, France

Popplewell, D.S.

AERE, National Radiological Protection Board,
Harwell, Didcot, Oxon OX11 0RA,
United Kingdom

Renzoni, A.

Università di Siena,
Via Cerchia 3, I-53100 Siena, Italy

Schulte, E.H.

CNEN-CEC,
I-19030 Fiascherino-La Spezia, Italy

Scoppa, P.

CNEN-CEC,
I-19030 Fiascherino-La Spezia, Italy

Thiels, G.M.*

CEC, Joint Research Centre, Chemistry Division,
I-21020 Ispra (Varese), Italy

Vanderborght, O.*

SCK/CEN,
Sektie Radionuclides Metabolism,
B-2400 Mol, Belgium

CHEMICAL METHODS OF DETERMINING
METAL FRACTIONS AND APPLICABILITY TO
TRANSURANICS ON FRESH, ESTUARINE AND COASTAL SEDIMENTS
(Session III)

Aston, S.R.*

Dept of Environmental Sciences,
The University of Lancaster,
Lancaster LA14YQ, United Kingdom

Baudin, G.

CEA, Centre d'études nucléaires de Fontenay-aux-Roses,
Service d'études analytiques,
B.P. 6, F-92260 Fontenay-aux-Roses, France

Bertozzi, G.

CEC, Joint Research Centre, Chemistry Division,
I-21020 Ispra (Varese), Italy

Boniforti, R.

CNEN, Laboratorio Studio Ambiente Marino,
I-19030 Fiascherino-La Spezia, Italy

Chester, R.*

Dept of Oceanography, The University of Liverpool,
P.O. Box 147, Liverpool L69 3BX, United Kingdom

Cigna, A.

CNEN-CSN Casaccia,
C.P. 2400, I-00100 Rome, Italy

Duursma, E.K.*

Delta Institute for Hydrobiological Research,
Vierstraat 28, N-4401 EA Yerseke, Netherlands

Frissel, M.J.*

Association Euratom-ITAL,
P.O. Box 48, N-6700 AA Wageningen, Netherlands

Guéguéniat, J.P.*

CEA/PSN/DPr, Laboratoire de radioécologie marine,
B.P. 270, F-50107 Cherbourg, France

Harvey, B.R.*

Ministry of Agriculture, Fisheries and Food,
Directorate of Fisheries Research,
Fisheries Radiobiological Laboratory,
Lowestoft, Suffolk NR33 0HT,
United Kingdom

MacKenzie, A.B.*

Scottish Universities Research and Reactor Centre,
East Kilbride, Glasgow G75 0QU,
United Kingdom

Mathew, E.*

Environmental Studies Section, Health Physics Division,
Bhabha Atomic Research Centre,
Trombay, Bombay 400 085, India

Murray, C.N.*

CEC, Joint Research Centre, Chemistry Division,
I-21020 Ispra (Varese), Italy

Rossi Cigna, L.

CNEN, Impianto EUREX,
I-13040 Saluggia (Vercelli), Italy

Saas, A.

CEA, Centre d'études nucléaires de Cadarache,
Département de Protection,
Laboratoire de radioécologie terrestre,
B.P. 1, F-13115 Saint-Paul-lez-Durance, France

---

* Presented Paper in Plenary Session.

The following conversion table is provided for the convenience of readers

FACTORS FOR CONVERTING SOME OF THE MORE COMMON UNITS TO INTERNATIONAL SYSTEM OF UNITS (SI) EQUIVALENTS

NOTES:
(1) SI base units are the metre (m), kilogram (kg), second (s), ampere (A), kelvin (K), candela (cd) and mole (mol).
(2) ▶ indicates SI derived units and those accepted for use with SI;
  ▷ indicates additional units accepted for use with SI for a limited time.
  [*For further information see the current edition of The International System of Units (SI), published in English by HMSO, London, and National Bureau of Standards, Washington, DC, and International Standards ISO-1000 and the several parts of ISO-31, published by ISO, Geneva.*]
(3) The correct symbol for the unit in column 1 is given in column 2.
(4) ✳ indicates conversion factors given exactly; other factors are given rounded, mostly to 4 significant figures:
  ≡ indicates a definition of an SI derived unit: [ ] in columns 3+4 enclose factors given for the sake of completeness.

Column 1 Multiply data given in:	Column 2	Column 3 by:	Column 4 to obtain data in:	
Radiation units				
▶ becquerel	1 Bq	(has dimensions of s^{-1})		
disintegrations per second (= dis/s)	$1 s^{-1}$	$\equiv 1.00 \times 10^0$	Bq	✳
▷ curie	1 Ci	$= 3.70 \times 10^{10}$	Bq	✳
▷ roentgen	1 R	$[= 2.58 \times 10^{-4}$	C/kg]	✳
▶ gray	1 Gy	$[\equiv 1.00 \times 10^0$	J/kg]	✳
▷ rad	1 rad	$= 1.00 \times 10^{-2}$	Gy	✳
▶ sievert *(radiation protection only)*	1 Sv	$[\equiv 1.00 \times 10^0$	J/kg]	✳
rem *(radiation protection only)*	1 rem	$[= 1.00 \times 10^{-2}$	J/kg]	✳
Mass				
▶ unified atomic mass unit ($\frac{1}{12}$ of the mass of ^{12}C)	1 u	$[= 1.660\,57 \times 10^{-27}$ kg, approx.]		
▶ tonne (= metric ton)	1 t	$[= 1.00 \times 10^3$	kg]	✳
pound mass (avoirdupois)	1 lbm	$= 4.536 \times 10^{-1}$	kg	
ounce mass (avoirdupois)	1 ozm	$= 2.835 \times 10^1$	g	
ton (long) (= 2240 lbm)	1 ton	$= 1.016 \times 10^3$	kg	
ton (short) (= 2000 lbm)	1 short ton	$= 9.072 \times 10^2$	kg	
Length				
statute mile	1 mile	$= 1.609 \times 10^0$	km	
nautical mile (international)	1 n mile	$= 1.852 \times 10^0$	km	✳
yard	1 yd	$= 9.144 \times 10^{-1}$	m	✳
foot	1 ft	$= 3.048 \times 10^{-1}$	m	✳
inch	1 in	$= 2.54 \times 10^1$	mm	✳
mil (= 10^{-3} in)	1 mil	$= 2.54 \times 10^{-2}$	mm	✳
Area				
▷ hectare	1 ha	$[= 1.00 \times 10^4$	m^2]	✳
▷ barn *(effective cross-section, nuclear physics)*	1 b	$[= 1.00 \times 10^{-28}$	m^2]	✳
square mile, (statute mile)^2	1 mile^2	$= 2.590 \times 10^0$	km^2	
acre	1 acre	$= 4.047 \times 10^3$	m^2	
square yard	1 yd^2	$= 8.361 \times 10^{-1}$	m^2	
square foot	1 ft^2	$= 9.290 \times 10^{-2}$	m^2	
square inch	1 in^2	$= 6.452 \times 10^2$	mm^2	
Volume				
▶ litre	1 l *or* 1 ltr	$[= 1.00 \times 10^{-3}$	m^3]	✳
cubic yard	1 yd^3	$= 7.646 \times 10^{-1}$	m^3	
cubic foot	1 ft^3	$= 2.832 \times 10^{-2}$	m^3	
cubic inch	1 in^3	$= 1.639 \times 10^4$	mm^3	
gallon (imperial)	1 gal (UK)	$= 4.546 \times 10^{-3}$	m^3	
gallon (US liquid)	1 gal (US)	$= 3.785 \times 10^{-3}$	m^3	

This table has been prepared by E.R.A. Beck for use by the Division of Publications of the IAEA. While every effort has been made to ensure accuracy, the Agency cannot be held responsible for errors arising from the use of this table.

Column 1 *Multiply data given in:*	Column 2	Column 3 *by:*	Column 4 *to obtain data in:*	
Velocity, acceleration				
foot per second (= fps)	1 ft/s	$= 3.048 \times 10^{-1}$	m/s	*
foot per minute	1 ft/min	$= 5.08 \times 10^{-3}$	m/s	*
mile per hour (= mph)	1 mile/h	$= \begin{cases} 4.470 \times 10^{-1} \\ 1.609 \times 10^{0} \end{cases}$	m/s km/h	
▷ knot (international)	1 knot	$= 1.852 \times 10^{0}$	km/h	*
free fall, standard, g		$= 9.807 \times 10^{0}$	m/s^2	
foot per second squared	$1\ ft/s^2$	$= 3.048 \times 10^{-1}$	m/s^2	*
Density, volumetric rate				
pound mass per cubic inch	$1\ lbm/in^3$	$= 2.768 \times 10^{4}$	kg/m^3	
pound mass per cubic foot	$1\ lbm/ft^3$	$= 1.602 \times 10^{1}$	kg/m^3	
cubic feet per second	$1\ ft^3/s$	$= 2.832 \times 10^{-2}$	m^3/s	
cubic feet per minute	$1\ ft^3/min$	$= 4.719 \times 10^{-4}$	m^3/s	
Force				
▶ newton	1 N	$[\equiv 1.00 \times 10^{0}$	$m \cdot kg \cdot s^{-2}]$	*
dyne	1 dyn	$= 1.00 \times 10^{-5}$	N	*
kilogram force (= kilopond (kp))	1 kgf	$= 9.807 \times 10^{0}$	N	
poundal	1 pdl	$= 1.383 \times 10^{-1}$	N	
pound force (avoirdupois)	1 lbf	$= 4.448 \times 10^{0}$	N	
ounce force (avoirdupois)	1 ozf	$= 2.780 \times 10^{-1}$	N	
Pressure, stress				
▶ pascal	1 Pa	$[\equiv 1.00 \times 10^{0}$	$N/m^2]$	*
▷ atmosphere[a], standard	1 atm	$= 1.013\ 25 \times 10^{5}$	Pa	*
▷ bar	1 bar	$= 1.00 \times 10^{5}$	Pa	*
centimetres of mercury (0°C)	1 cmHg	$= 1.333 \times 10^{3}$	Pa	
dyne per square centimetre	$1\ dyn/cm^2$	$= 1.00 \times 10^{-1}$	Pa	*
feet of water (4°C)	$1\ ftH_2O$	$= 2.989 \times 10^{3}$	Pa	
inches of mercury (0°C)	1 inHg	$= 3.386 \times 10^{3}$	Pa	
inches of water (4°C)	$1\ inH_2O$	$= 2.491 \times 10^{2}$	Pa	
kilogram force per square centimetre	$1\ kgf/cm^2$	$= 9.807 \times 10^{4}$	Pa	
pound force per square foot	$1\ lbf/ft^2$	$= 4.788 \times 10^{1}$	Pa	
pound force per square inch (= psi)[b]	$1\ lbf/in^2$	$= 6.895 \times 10^{3}$	Pa	
torr (0°C) (= mmHg)	1 torr	$= 1.333 \times 10^{2}$	Pa	
Energy, work, quantity of heat				
▶ joule ($\equiv W \cdot s$)	1 J	$[\equiv 1.00 \times 10^{0}$	$N \cdot m]$	*
▶ electronvolt	1 eV	$[= 1.602\ 19 \times 10^{-19}$	J, approx.]	
British thermal unit (International Table)	1 Btu	$= 1.055 \times 10^{3}$	J	
calorie (thermochemical)	1 cal	$= 4.184 \times 10^{0}$	J	*
calorie (International Table)	$1\ cal_{IT}$	$= 4.187 \times 10^{0}$	J	
erg	1 erg	$= 1.00 \times 10^{-7}$	J	*
foot-pound force	$1\ ft \cdot lbf$	$= 1.356 \times 10^{0}$	J	
kilowatt-hour	$1\ kW \cdot h$	$= 3.60 \times 10^{6}$	J	*
kiloton explosive yield (PNE) ($\equiv 10^{12}$ g-cal)	1 kt yield	$\simeq 4.2 \times 10^{12}$	J	

[a] atm (g) (= atü): atmospheres gauge
atm abs (= ata): atmospheres absolute

[b] lbf/in^2 (g) (= psig): gauge pressure;
lbf/in^2 abs (= psia): absolute pressure.

Column 1 *Multiply data given in:*	Column 2	Column 3 *by:*	Column 4 *to obtain data in:*	

Power, radiant flux

▶ watt — 1 W — $[\equiv 1.00 \times 10^0$ — J/s] — *

British thermal unit (International Table) per second — 1 Btu/s — $= 1.055 \times 10^3$ — W

calorie (International Table) per second — 1 cal$_{IT}$/s — $= 4.187 \times 10^0$ — W

foot-pound force/second — 1 ft·lbf/s — $= 1.356 \times 10^0$ — W

horsepower (electric) — 1 hp — $= 7.46 \times 10^2$ — W — *

horsepower (metric) (= ps) — 1 ps — $= 7.355 \times 10^2$ — W

horsepower (550 ft·lbf/s) — 1 hp — $= 7.457 \times 10^2$ — W

Temperature

▶ kelvin — K

▶ degrees Celsius, t — $t = T - T_0$ — *
    where T is the thermodynamic temperature in kelvin
    and T_0 is defined as 273.15 K

degree Fahrenheit — $t_{°F} - 32$
degree Rankine — $T_{°R}$ $\left. \right\} \times \left(\dfrac{5}{9}\right)$ gives $\left\{ \begin{array}{l} t \ \textit{(in degrees Celsius)} \ * \\ T \ \textit{(in kelvin)} \qquad\qquad * \\ \Delta T \ (= \Delta t) \qquad\qquad\quad * \end{array} \right.$
temperature difference^c — $\Delta T_{°R} \ (= \Delta t_{°F})$

Thermal conductivity ^{c}

1 Btu·in/(ft^2·s·°F)   *(International Table Btu)* $= 5.192 \times 10^2$ $W·m^{-1}·K^{-1}$

1 Btu/(ft·s·°F)   *(International Table Btu)* $= 6.231 \times 10^3$ $W·m^{-1}·K^{-1}$

1 cal$_{IT}$/(cm·s·°C) $= 4.187 \times 10^2$ $W·m^{-1}·K^{-1}$

Miscellaneous quantities

litre per mole per centimetre $(1M/cm =) \ 1 \ ltr·mol^{-1}·cm^{-1}$ $= 1.00 \times 10^{-1} \ m^2/mol$ *
(molar extinction coefficient or molar absorption coefficient)

G-value, traditionally quoted per 100 eV
    of energy absorbed $1 \times 10^{-2} \ eV^{-1} = 6.24 \times 10^{16} \ J^{-1}$
(radiation yield of a chemical substance)

mass per unit area $1 \ g/cm^2$ $[= 1.00 \times 10^1 \ kg/m^2]$ *
(absorber thickness and mean mass range)

---

^{c} A temperature interval or a Celsius temperature difference can be expressed in degrees Celsius as well as in kelvins.

HOW TO ORDER IAEA PUBLICATIONS

An exclusive sales agent for IAEA publications, to whom all orders
and inquiries should be addressed, has been appointed
in the following country:

UNITED STATES OF AMERICA   UNIPUB, 345 Park Avenue South, New York, NY 10010

In the following countries IAEA publications may be purchased from the
sales agents or booksellers listed or through your
major local booksellers. Payment can be made in local
currency or with UNESCO coupons.

ARGENTINA	Comisión Nacional de Energía Atomica, Avenida del Libertador 8250, RA-1429 Buenos Aires
AUSTRALIA	Hunter Publications, 58 A Gipps Street, Collingwood, Victoria 3066
BELGIUM	Service Courrier UNESCO, 202, Avenue du Roi, B-1060 Brussels
CZECHOSLOVAKIA	S.N.T.L., Spálená 51, CS-113 02 Prague 1
	Alfa, Publishers, Hurbanovo námestie 6, CS-893 31 Bratislava
FRANCE	Office International de Documentation et Librairie, 48, rue Gay-Lussac, F-75240 Paris Cedex 05
HUNGARY	Kultura, Hungarian Foreign Trading Company P.O. Box 149, H-1389 Budapest 62
INDIA	Oxford Book and Stationery Co., 17, Park Street, Calcutta-700 016
	Oxford Book and Stationery Co., Scindia House, New Delhi-110 001
ISRAEL	Heiliger and Co., Ltd., Scientific and Medical Books, 3, Nathan Strauss Street, Jerusalem 94227
ITALY	Libreria Scientifica, Dott. Lucio de Biasio "aeiou", Via Meravigli 16, I-20123 Milan
JAPAN	Maruzen Company, Ltd., P.O. Box 5050, 100-31 Tokyo International
NETHERLANDS	Martinus Nijhoff B.V., Booksellers, Lange Voorhout 9-11, P.O. Box 269, NL-2501 The Hague
PAKISTAN	Mirza Book Agency, 65, Shahrah Quaid-e-Azam, P.O. Box 729, Lahore 3
POLAND	Ars Polona-Ruch, Centrala Handlu Zagranicznego, Krakowskie Przedmiescie 7, PL-00-068 Warsaw
ROMANIA	Ilexim, P.O. Box 136-137, Bucarest
SOUTH AFRICA	Van Schaik's Bookstore (Pty) Ltd., Libri Building, Church Street, P.O. Box 724, Pretoria 0001
SPAIN	Diaz de Santos, Lagasca 95, Madrid-6
	Diaz de Santos, Balmes 417, Barcelona-6
SWEDEN	AB C.E. Fritzes Kungl. Hovbokhandel, Fredsgatan 2, P.O. Box 16356, S-103 27 Stockholm
UNITED KINGDOM	Her Majesty's Stationery Office, Agency Section PDIB, P.O. Box 569, London SE1 9NH
U.S.S.R.	Mezhdunarodnaya Kniga, Smolenskaya-Sennaya 32-34, Moscow G-200
YUGOSLAVIA	Jugoslovenska Knjiga, Terazije 27, P.O. Box 36, YU-11001 Belgrade

---

Orders from countries where sales agents have not yet been appointed and
requests for information should be addressed directly to:

Division of Publications
International Atomic Energy Agency
Wagramerstrasse 5, P.O. Box 100, A-1400 Vienna, Austria